Sexual Pedagogies

Sexual Pedagogies

Sex Education in Britain, Australia, and America, 1879–2000

Edited by
Claudia Nelson and Michelle H. Martin

palgrave
macmillan

SEXUAL PEDAGOGIES

Softcover reprint of the hardcover 1st edition 2004 978-1-4039-6350-5

First published 2004 by
PALGRAVE MACMILLAN™
175 Fifth Avenue, New York, N.Y. 10010 and
Houndmills, Basingstoke, Hampshire, England RG21 6XS
Companies and representatives throughout the world

PALGRAVE MACMILLAN is the global academic imprint of the Palgrave Macmillan division of St. Martin's Press, LLC and of Palgrave Macmillan Ltd. Macmillan® is a registered trademark in the United States, United Kingdom and other countries. Palgrave is a registered trademark in the European Union and other countries.

ISBN 978-1-349-52752-6 ISBN 978-1-4039-8103-5 (eBook)
DOI 10.1057/9781403981035

Library of Congress Cataloging-in-Publication Data
Sexual pedagogies: sex education in Britain, Australia, and America, 1879–2000 / edited by Claudia Nelson and Michelle H. Martin.
p. cm.
Includes bibliographical references and index.

1. Sex instruction—Great Britain—History. 2. Sex instruction—Australia—History. 3. Sex instruction—United States—History. 4. Sex instruction literature—Great Britain—History. 5. Sex instruction literature—Australia—History. 6. Sex instruction literature—United States—History. I. Nelson, Claudia. II. Martin, Michelle H., 1966–

HQ57.6G7S49 2003
613.9'071—dc21 20033050575

A catalogue record for this book is available from the British Library.

This book is printed on paper suitable for recycling and made from fully managed and sustained forest sources. Logging, pulping and manufacturing processes are expected to conform to the environmental regulations of the country of origin.

Design by Newgen Imaging Systems (P) Ltd., Chennai, India.

First edition: January 2004
10 9 8 7 6 5 4 3 2 1

To my supportive spouse, Glenn, who is sometimes more of a feminist than I.

—*M.H.M.*

To Anne, for advice and consent.

—*C.N.*

CONTENTS

Acknowledgments

Both of us wish to thank Jack Zipes for his support of this anthology; Claudia Nelson also thanks Angela Ingram for her valuable suggestions in the planning stage, and Texas State University, San Marcos for providing grant support to help underwrite the collection's completion. In addition, we extend our appreciation to the Children's Literature Association for its continuing encouragement of the scholarly pursuit of children's and adolescent studies and for creating a space where interdisciplinary projects such as this one can flourish. Finally we are deeply grateful to our contributors for their patience, hard work, and commitment to excellence.

Contributors

Lesley Hall is archivist at the Wellcome Library for the History and Understanding of Medicine, London, and an honorary lecturer in History of Medicine, University College London. She has published extensively on sex and gender in nineteenth- and twentieth-century Britain. Current projects include a biography of feminist sex radical Stella Browne and a survey of British women writing about sexuality, ca. 1860–1960.

Associate Professor *Margot Hillel* is head of the School of Arts and Sciences (Victoria) at Australian Catholic University. She has published widely in the field of children's literature, both in scholarly and professional journals and in textbooks for teachers, and is president of the Australasian Children's Literature Association for Research.

Kenneth Kidd is assistant professor of English at the University of Florida, where he teaches courses in children's literature and film and serves as Associate Graduate Coordinator. His book *Making American Boys: The Science of Boyology and the Feral Tale* is forthcoming from the University of Minnesota Press.

Michelle Martin is assistant professor of English at Clemson University in South Carolina. In addition to her research interests in menstruation and its connections with children's literature, she has interests in African American children's picture books. Her book *Brown Gold: Milestones of African American Children's Picture Books, 1845–2002*, is forthcoming from Routledge in 2004. Because of a strong belief in using her academic pursuits to make a difference in the community, she has hosted two Harry Potter Parties, which 1,400 Clemson children and adults attended in 2001 and 2002, and gives away thousands of free books to children in the community every year.

Claudia Nelson is associate professor of English at Texas A&M University. She is the author of *Boys Will Be Girls: The Feminine Ethic and British Children's Fiction, 1857–1917* (1991), *Invisible Men: Fatherhood in Victorian Periodicals, 1850–1910* (1995), and *Little*

Strangers: Portrayals of Adoption and Foster Care in America, 1851–1929 (2003). She has also coedited two earlier anthologies, *The Girl's Own: Cultural Histories of the Anglo-American Girl, 1830–1915* (with Lynne Vallone) and *Maternal Instincts: Visions of Motherhood and Sexuality in Britain, 1875–1925* (with Ann Sumner Holmes).

Sharyn Pearce is senior lecturer in the Faculty of Creative Industries at the Queensland University of Technology in Brisbane, Australia. She has published extensively in the areas of children's literature and Australian studies, and has recently edited (with Kerry Mallan) *Youth Cultures: Texts, Images and Identities* (Greenwood, 2003).

Angela M. Salas received her Ph.D. from the University of Nebraska in 1995. She subsequently taught at Adrian College, where she served as honors chair, department chair, and faculty president, in addition to teaching composition and literature to undergraduate students. Salas has published articles on Yusef Komunyakaa, Willa Cather, and Edith Wharton. Her first book, *Flashback through the Heart: The Poetry of Yusef Komunyakaa*, is forthcoming from Susquehanna University Press.

Associate Professor *Shurlee Swain* is a reader in History at Australian Catholic University and a senior research fellow in the Department of History, University of Melbourne. She has researched widely in the history of women and children, her latest book being *Confronting Cruelty: Historical Perspectives on Child Protection in Australia* (Melbourne University Press, 2002).

Ellen Warne is research fellow in the Department of History at the University of Melbourne. Her doctoral research focused on the conservative feminism of Christian women's organizations in Australia between 1890 and 1940. She is currently working on a project with Patricia Grimshaw and Shurlee Swain on working mothers in Australian history.

Introduction

Claudia Nelson and Michelle H. Martin

If, as the feminist catchphrase has it, "the personal is political," then sex education—surely a thoroughly personal matter—is thoroughly political as well. From the 1960s onward, Americans, like Australians and Britons, have often seen sex education's political aspects as a matter of party affiliation (see Moran). Children are typically cast as pawns in these culture wars, which show no signs of diminishing; for example, when Dr. David Satcher, appointed surgeon general of the United States by President Clinton, issued under the new Republican administration a June 2001 report entitled "The Call to Action to Promote Sexual Health and Responsible Sexual Behavior" urging "equity of opportunity for sex education" through detailed classroom instruction, many conservatives were predictably outraged and many liberals, equally predictably, gratified (see, e.g., Schemo). In the United States, at least, the dominant issue in debates such as the one surrounding Satcher's report often becomes who should serve as children's ultimate moral authority. Are social mores, and particularly sexual mores, to be inculcated by schools or by families? How much freedom should parents have—for we often assume that children should have little—to question the government's decisions about what kind of education serves the public good? How much freedom should school nurses and teachers have to cater to the sex-education needs of their particular schools? Since a great many political debates deal with how much power the ordinary individual should have, sex education may come to look like one of many planks in party platforms, something to be discussed alongside gun control or tax relief.

But the evolving discourse that is sex education is not limited to pundits—or to health professionals or teachers, for that matter. For one thing, we have the question of what the term "sex education" may be said to include. What kinds of knowledge are being passed along? Who is imparting this knowledge, and who receiving it? Where does the transmittal of information take place? What teaching aids are in use, and what are understood to be the consequences of students' learning or refusing to learn? Change enough of these variables, and we may find ourselves far from the kind of sex education that Satcher clearly has in mind, even while one participant is undeniably still teaching another something about human sexuality.

Thus "sex education" may refer to a parental explanation designed to enlighten a small child about the birth process, or to a school class for preteens that outlines methods of contraception and the avoidance of sexually transmitted diseases. But the same term may also reasonably be used to describe sexually didactic children's books or "teenpix," marriage manuals or other materials calculated to encourage adult sexual fulfillment, cautionary films about venereal disease aimed at soldiers in the First or Second World Wars, or even magazine articles outlining the principles behind human cloning or the latest infertility treatments. As Jeffrey P. Moran points out, sex education is not a stable entity, but something that quickly responds to national crises or to changes in the social ethos.[1] It reflects evolving ideas about gender, race, social class, and childhood as well as about sexuality. It is also not a new phenomenon; for example, in *The Facts of Life*, Roy Porter and Lesley Hall begin their discussion of British sex instruction with the year 1650. Perhaps we should speak not of sex education but of sex educations.

Not surprisingly, then, the chapters contained in the present book do not give a comprehensive picture of the forums, mediums, and ideologies influencing the sexual curricula of three intertwined societies—Britain, Australia, and the United States—over 120 years; the subject is too broad to be covered thoroughly in one modest volume. Rather, it is our project to suggest *some* of the conflicting ways in which sex education has functioned during this period, both for children and for adults. Even this slim collection chronicles many of these functions: to instill moral and physical self-control but also to remove inhibitions; to help contain the sexual energies of the young, the poor, or the racial Other but also to profit from them; to advance but also to retard the cause of feminism; to preach both tolerance and intolerance of sexual minorities; to disguise the marketing of commercial products and gender-role stereotypes alike. Moran has recently presented considerable evidence suggesting that sex education, at least in its more obvious forms, may have little effect on its consumers. But whether or not he is correct, clearly such instruction is commonly *perceived* as a powerful and flexible social tool, which is why so many politicians have been interested in it. Certainly it is concerned with sexuality, but it is concerned with many other aspects of our lives as well.

One idea animating the array of chapters here, in short, is that "sex education" should be broadly defined. As professional sex educators (or commentators on professional sex education) often note, people get information about sex from all kinds of places—and almost all of these sources will attract strong criticism from someone. Indeed, we may see a tendency to take the word for the deed, so common is the implication that "wrong" teaching about sex is itself a source of sexual corruption. At the turn of the twentieth century, for instance, Anglophone writers sometimes warned not only that children should avoid getting the "facts of life" from their peers or their servants but also that parental advice, too, might be inadequate or unscientific. As these writers saw it, the ill-informed parent would leave the child ignorant and vulnerable, while the too-knowledgeable schoolmate or hireling would encourage overconfidence by focusing on sex's pleasures rather than its potential risks. Moreover, as Patty Campbell observes in her study *Sex Guides*, historically authors of manuals for teenagers have "often denounced their predecessors' work as worthless and even vicious" (4).

But given the expert's traditional impatience with divergent points of view, this struggle for authority over the student is not limited to writers for children. Thus the Dutch gynecologist Theodoor Van de Velde spends considerable time in his 1926 volume *Ideal Marriage* criticizing Marie Stopes's 1918 *Married Love* (see, e.g., 202–04, 285–88); one sign of the changing times, however, is that Van de Velde worries that following Stopes's advice will be risky because couples will not have a sufficiently active sex life.[2] And even when the official and/or credentialed discourse seems unified and consistent, it must always operate in the context of any number of unofficial discourses, many of which will undermine the "authoritative" point of view. A case in point here is the underground magazine *Out of Bounds*, produced by British public-school students and dropouts in the mid-1930s. Its second number repeatedly voices complaints about the educational system's desire to preserve pupils' ostensible innocence instead of providing what coeditor Giles Romilly calls "intelligent [sex] instruction" (34); as one contributor, Gavin Ewart, observes, the official position has counterproductive results:

> "Sex is God's and you mustn't touch it,
> It's a beautiful shoe whose very latchet
> You may not loose."
>
> So we were onanists. . . . (12, lines 35–37)

The rebellious stance of these teenage writers reminds us that the providers (or nonproviders) of sex education are not the only ones to hold opinions on the effects that the presence or absence of such instruction may have. Consumers (or would-be consumers) may have strong views too. In short, to consider only the official discourse is to take the tree for the forest.

The other major idea shaping the construction of this anthology is the perception that sex education is appropriately viewed in an international context. Britain, the United States, and Australia are a case in point. Their shared language and heritage have encouraged an ongoing cultural cross-pollination. In the late nineteenth and early twentieth centuries, the era when formal instruction in human sexuality for a general audience took hold in our society and the point at which this book begins, such cross-pollination was furthered by the veneration that Britain commanded from many Americans and Australians; Daniel Rodgers has chronicled in his 1998 study *Atlantic Crossings: Social Politics in a Progressive Age* the tendency of American reformers to look to Europe, and particularly to England, for guidance, while Australia's colonial status ensured that it would import many of its social beliefs from the mother country. Thus, for example, Shurlee Swain, Ellen Warne, and Margot Hillel discuss in the present volume the pre–First World War calls for a sexual pedagogy dominated by schoolteachers, clerics, and physicians, and the subsequent competition among these professional groups over which should have the greatest authority. Swain, Warne, and Hillel focus on this debate in an Australian setting, but the same discourse was a feature of life in 1890s Britain (see, e.g., Nelson, "Guidance" and "Patients"). Later, as Sharyn Pearce notes in her contribution to this anthology, influence patterns shifted toward the United States (while the influence itself remained a constant) under stimuli such as American soldiers' presence in both Britain and Australia during the Second World War. In addition, of course, the ease with which magazines, books, or movies—and now, E-mail and documents posted on the Internet—may circulate among the three nations remains an important factor.

But from the 1880s through the end of the 1920s, the mutual support and reciprocal influence that we may see among Australia, the United States, and Britain (and other Anglophone countries not discussed here) were also spurred by the comparatively small size of the community of sex educators.[3] To be sure, this community had its distinct subdivisions. In one corner of the field we might place scientifically oriented sex researchers writing for an adult audience after the fashion of Havelock Ellis, English author of the magisterial *Studies in the Psychology of Sex* (7 vols., 1896–1928), whose predecessors and closest ideological allies were Continental and whose most information-packed work was subject to rigorous censorship by the authorities of Britain, America, and Australia alike. Another corner might belong to authors of sexually didactic (and, often, erotic) adult fiction, whose writings were also officially discouraged—especially upon the emergence of a sufficiently energetic public figure, such as Anthony Comstock of the New York Society for the Suppression of Vice or Sir William Joynson-Hicks, the British home secretary known as "The Policeman of the Lord." Some of these writers saw themselves as beleaguered crusaders on an international level, notably Radclyffe Hall, whose pioneering and polemical lesbian novel *The Well of Loneliness* (1928) was banned as obscene in its native Britain and in Australia but judged publishable, after a court battle, in the United States. Before the 1920s, however, much of the fiction designed principally to illuminate sexuality lacked *The Well*'s claims to literary worth, belonging instead to the thriving—and, again, multinational—late-Victorian pornography industry. Nevertheless, much as Hall's novel stressed its educational function by including a prefatory note by Ellis, Victorian erotica might also frame itself as instruction, a phenomenon that Claudia Nelson's chapter on *The Pearl* explores in this volume.

Purity reformers made up still a third division of early sex educators, one that differed from the other two in its greater interest in children and that saw itself, and rightly so, as constituting what was in effect a political pressure group. As such, this group shared with the sexologists the tendency to form alliances with like-minded foreigners, who could offer inspiration, advice, and educational materials to be reprinted and distributed in the new locale. Purity organizations, founded to combat social problems such as prostitution and venereal disease by encouraging the adoption of a single (female) sexual standard and the publicizing of certain kinds of sexual information among young women, had chapters worldwide and helped to shape the sexual pedagogy of both printed manuals and classroom instruction for many decades. Thus, for example, the White Cross Society, begun in Britain in the 1880s, spread not only to Australia, the United States, and other English-speaking nations but also to the European continent. Similarly, Warne has chronicled elsewhere the importance of an organization native to the United States, the Women's Christian Temperance Union (WCTU), and two groups founded in Britain—the Young Women's Christian Association (YWCA) and the Mothers' Union—in the effort "to establish community-based sex education taught by women" for Australian youngsters between 1891, when the WCTU reached that country, and the early

1930s (311). The model of "informed innocence," discussed here in an Australian context by Swain, Warne, and Hillel, has an extensive international pedigree. We might note that American purity workers, too, readily accepted "foreign" purity literature for local distribution, since ideologically it was not foreign at all. Thus turn-of-the-century American pamphlets such as Dr. L. Anna Ballard's "Danger to Our Girls," a plea to mothers for better sex education, typically included advertisements of tracts by both American and British authors, with no distinctions of nationality made.[4]

Sex educators have also long formed personal ties internationally, another circumstance that helps to show where the ideological lines might be drawn. For instance, Dr. Mary Wood-Allen (d. 1908), an American author of numerous instructional manuals for the young who became World Superintendent of the WCTU's Purity Department, was in communication with British purity leader Josephine Butler in the 1890s. (Wood-Allen had connections to Australia as well, in that one of her pamphlets was the basis for a 1918 sex manual created by the Anglican bishops of Tasmania and Victoria.[5]) Anti-vice crusader William T. Stead, a British journalist who famously warned against the sexual ignorance (and hence vulnerability) of working-class girls in his series of articles "The Maiden Tribute of Modern Babylon" (1885), was an enthusiastic writer of American-themed books and pamphlets who had many friends and admirers in that country. And by the 1920s, when the White Cross dictum "Keep thyself pure" was beginning to give way to the modernist emphasis on scientific family planning, Australian eugenist, birth-control advocate, and sex educator Marion Piddington had begun a twenty-one-year correspondence with her British colleague Stopes, who in turn had become acquainted with Margaret Sanger when the response of the American postal authorities to Sanger's periodical *The Woman Rebel* forced the latter to relocate temporarily to England in 1914 (Wyndham 74, Campbell 73).[6] These few examples are representative; despite the great distance separating British, Australian, and American sex educators and the inevitable professional jealousies that may be found, from the 1880s onward the community's various factions have been in many ways tightly knit.

There are, of course, many other instances of the multinational scope of what Sue Morgan has termed "agenda[s] for sexual political change" (19). More recent phenomena of this type include the educational efforts related to AIDS prevention, which often take the form of world AIDS conferences, statements made under the aegis of the United Nations, and so on. The activist AIDS group ACT UP, founded in New

York City in March 1987, established chapters in London in 1989 and Sydney in 1990. As ACT UP's spread within and beyond English-speaking nations suggests, the international discourse surrounding AIDS education has from the beginning been markedly politicized, a circumstance that has continued even as world attention gradually moves away from gay men in Western nations and toward heterosexuals in sub-Saharan Africa. Moreover, the latter shift has also spotlighted the role of market forces within sex education, as commentators ponder how to teach safe sex to populations who may consider condoms too expensive to use.

And beyond the realm of public health, economic realities fuel other forms of sex education as well. As the contributions to this volume by Lesley Hall, Michelle Martin, Kenneth Kidd, and Angela M. Salas suggest, sex education can bring considerable financial profit if appropriately manipulated—consider the royalties earned by British sex guru Alex Comfort when *The Joy of Sex* (1972) became an international bestseller, for example—but some didactic messages are evidently more salable than others. Thus Hall's exploration of the sexual guidance that British women received from authorities from Stopes onward indicates that women have historically been willing to buy magazines and marriage manuals presenting a picture of female sexuality that may bear little resemblance to many women's lived experience. Advice columns, Hall finds, often stressed sexual fulfillment even as mid-twentieth-century women continued to report the same ignorance of sex (and its potential consequences) that had horrified reformers such as Stead and his allies in the 1880s.

One might note here that Britain is not alone in producing women's magazines that have served as important vectors for sex education or, potentially, for pressuring readers into accepting an ideology of sexual liberation. Swain, Warne, and Hillel cite sex-education columns that began in an Australian journal, *Woman's Voice*, as early as 1895, while by 1910, the American *Ladies' Home Journal* was publishing a similar column entitled—perhaps in imitation of *Woman's Voice*'s "Shall We Tell the Children?"—"What Shall We Tell the Children?", run by Mary Wood-Allen's daughter, Rose Woodallen Chapman. More recently, periodicals such as *Cosmopolitan* have relentlessly sold a particular sexual role model, one characterized by her encyclopedic knowledge (gained, we are to suppose, from careful study of *Cosmopolitan* and its fellows rather than from instinct) of how best to please men in bed. If the patterns that Hall traces hold true for the United States as well, however, we might hypothesize that women are much more likely to buy the "Cosmo girl" than to become her. Sex instruction sells, but it does not always transform its readers.

If some magazines find it feasible and even profitable to participate in a liberationist effort to enact "sexual political change" because they are perceived as targeting an adult audience, products aimed at younger consumers may well feel social and thus economic pressure to take a comparatively conservative stance.[7] As Campbell provocatively remarks, "the one [stable] characteristic of sex education for young people in America from Victorian to modern times has been the reluctance of sex educators, other professionals and parents to tell teenagers what they really need to know about sex" (3). Moreover, because materials for the sex instruction of children are often reprinted (with or without editorial updating) over several decades, even those texts that might have looked radical at their debut may seem comfortably reactionary a generation later.[8] In examining pamphlets distributed from the 1940s onward at school showings of American puberty-education films sponsored by manufacturers of "feminine hygiene" products, Martin notes a similar tendency to recycle content and (conservative) ideology, although style is subject to continual updating in an effort to connect with potential consumers.

Texts for children that propose to entertain their constituents in addition to instructing them offer additional examples of the need for conservatism. While Robert Eberwein has focused on the tendency of films intended for classroom use to "narrativize the act of medical, scientific and pedagogic vision" (5), Kidd conversely suggests that films intended for theatrical release may hearken back to pedagogical ancestors in their use of voiceovers, among other traits. He sees parallels between the ideological content of representative popular Hollywood movies for teens from the 1980s and 1990s and that of much earlier instructional films on interacting with one's peers, a subject that frequently forms part of the curriculum in school sex-education courses and teen sex guides (see, e.g., Campbell's chapter on the 1950s, 107–33). Making movies is expensive, of course, and accordingly studios tend to rely on tried-and-true formulas and to attempt to satisfy not only the iconoclastic teen viewer in New York or Los Angeles but also, say, the film censorship board in Australia. The result, Kidd notes, is often a conservative film with a veneer of sexual radicalism.

We may see a similar phenomenon in the recent gay-themed novels for young adults that Salas discusses, which offset their message of liberal tolerance with plot twists suggesting that adjustment to homosexuality (whether one's own or another's) is a battle rife with tragedy and alienation if not with outright violence. As Campbell's 1986 book observes, "Sex education in the form of a novel is subject to public attack and censorship attempts to a degree that most sex manuals, with their aura of clinical respectability, have never been (until recently)" (217); the 1990s YA novels examined in Salas's chapter appear to have been written by authors fully conscious of this situation. It is worth noting that the more radical picture books against which Salas sets these novels emanate, unlike the latter, from small presses founded, in part, to further the cause of gay rights, and hence presumably more willing than the large publishers to take certain kinds of risks. And we may also find it suggestive that while the comparatively conservative YA titles have taken root internationally—Salas looks at a lesbian novel by New Zealand author Paula Boock, for instance, and British entries in the field include work by writers such as Aidan Chambers—searching British and Australian online bookstores for gay-themed picture books is still likely to turn up only imported copies of *Heather Has Two Mommies* and its fellows.

Nevertheless, *Heather is* available in Britain and Australia, just as sexually didactic American films have circulated in both countries since the early silent period. And while these importations may arouse protest (Australian filmmakers and culture guardians have been complaining about what one 1935 pamphlet christened "Mental Rubbish from Overseas" since at least the mid-1920s, often citing the sexual content of the "rubbish" and its presumed bad effects upon the young; see, e.g., G. Turner), they show no signs of abating. But the cultural overlapping of sex instruction in all three nations is the product not of imported texts alone, but of shared ideas as well. Thus the late-nineteenth-century erotic magazine *The Pearl* had only a brief run in its native Britain, and the chances that it caught the eye of Australian children's novelist Ethel Turner are minute; even so, as the chapters by Nelson and Swain, Warne, and Hillel suggest, both *The Pearl* and Turner respond in their separate ways to a widespread tendency within prewar Anglophone society to sexualize the child. Similarly, the 1950s Australian sex-education manuals for Catholic and Protestant boys that Pearce's contribution examines share the longing for social stability, the respect for authority, and even the failed attempts at teen slang present in the American materials from the same era examined elsewhere by Campbell and Moran. And the marketing strategies that Martin discusses on the part of American sanitary-pad manufacturers have, as Hall notes, been employed in Britain as well.

In their various ways, all the chapters of this book look at points at which certain competing visions of sexuality intersect. Indeed, this phenomenon seems inseparable from any discussion of sex education, which must always presuppose the meeting of an "innocent" (or "ignorant") party with an "experienced" (or "knowledgeable," or even "depraved") one. But our discussions here are inflected by considerations that vary with each text examined, such as the motivation of the writer, the presumed age of the audience, and the degree of cultural authority granted to the genre. Some of these texts may be described as enjoying official sanction, for example because they emanated from governmental sources, bore the imprimatur of the religious establishment, or served as teaching tools in schools. Others, such as Hollywood films for young adults or advice columns in women's magazines, belong to the respectable segment of popular culture; still others drew their power from their outlaw status as emanations from the publishing underground. In addition, time is an important factor here, since in the period that this anthology addresses, understandings of sexuality changed dramatically in response to pressures including the medical "invention" of homosexuality, the gay rights movement, feminism's redefinition of femininity, pandemics of venereal disease, and the idealizing of sexual activities of all kinds.

But whether a given text was originally intended primarily as science or as entertainment, all these discourses—like all pedagogy, arguably—may be seen as exercises in social engineering. These are texts that are trying to influence the behavior of their intended audience, whether they seek to inspire loyalty to a particular brand of tampon, to encourage resistance to dominant understandings of sexuality, or to slow the pace of social change where gender roles are concerned. And since nowadays we frequently (if recklessly) assume that a text that tries to mold its public must be a text aimed at children, and that sexual ignorance is a characteristic limited to the young, we may tend to see sex education as something we soon outgrow.

This anthology posits that such perceptions are incomplete. We cannot afford them either as citizens or as investigators of the culture to which we belong. By limiting the scope of our understanding of sex education, we may miss what the differences among separate examples of sex instruction may have to reveal about the surrounding society. For one, we may notice that while texts aimed at young people seem calculated to discourage them from acting on their sexuality, texts aimed at adults not only take a far more permissive approach where adult behavior is concerned but also may celebrate the same youthful eroticism that the texts for children lament. If the teen films that Kidd examines guide boys to become their (heterosexual, middle-class) fathers, for instance, the adult comedy *It's in the Water* (1996) traces its married heroine's belated rebellion against the sterile values of her socialite mother, a rebellion that is enabled both by the protagonist's loyalty to the friends that she acquired in high school and by her renting of a clutch of films about teenage lesbians: viewing

The Incredibly True Story of Two Girls in Love and its ilk is apparently not only educational but also emancipatory. Thus examining children's and adults' sex literature side by side permits us to see the extent to which adults may appear to define the young as a distinct species, variously suspect, enviable, and desirable because of its peculiar relationship to sexuality. Whether written for adults or for children, texts about a young person's initiation into sex tend to suggest that this process is potentially dangerous, thrilling, and life-altering, all at the same time, and that this sexual melodrama is something that the adult may well need to recapture—perhaps even by writing sex guides—in order to remain fulfilled.

Notes

1. Indeed, Patty Campbell holds that the sudden appearance of sex guides aimed specifically at young readers, which she dates to the 1890s, was a response to "the signs of a disintegration of the old social order"; authors used this new form to "preserv[e] the sexual attitudes that had been one of [that order's] foundations" (15).
2. In its English translation (made originally by British sex radical Stella Browne), Van de Velde's marriage manual was a major success in terms of both sales figures and longevity. Some 700,000 copies had been printed by the early 1970s, more than 500,000 of them in the United States.
3. Arguably, the spokespersons for sexuality itself—or at least for fertility—outnumbered them. These years were the heyday of eugenics and the doctrine of the propagation of the fit (usually defined as the healthy White middle class), and the United States, Britain, and Australia all witnessed much anxiety about the low birth rate among the educated.
4. Many of the British pamphlets recommended at the back of "Danger to Our Girls" were written by Ellice Hopkins, cofounder of the White Cross Society. Ballard's comment, "The highest human faculty is that of sex in its mental and physical powers" (8), like her other arguments in this pamphlet, has much in common with the beliefs of Hopkins, who combined purity work with an "exaltation of marital sexuality" (Morgan 26).
5. The Women's Library of London Guildhall University holds two letters from Wood-Allen to Butler, both dated 1897. For the transformation of Wood-Allen's *In Her Teens* into *Helps to Parents in Explaining Matters of Sex to the Young*, see Warne 316 and Warne, Swain, and Hillel in this volume. Australia continued to import many of its sex-education manuals; later, for instance, the New South Wales Department of Education would adopt texts by Cyril Bibby, British editor of the *International Journal of Sexology* in the years immediately following the Second World War (Wyndham 163, fn. 907).
6. In addition to their other activities, Piddington, Sanger, and Stopes all wrote sex-instruction manuals that met with official disapproval; this shared experience may also have strengthened the close bond that Sanger formed with Ellis during her London exile. In the late 1920s, Stopes's *Married Love* was

still being cited on the floor of the Irish parliament as an example of the kind of book that should not be allowed into the country.

7. To be sure, this pressure may work in both directions, perhaps influenced by the importance granted to youth at any given moment. As historian of sexuality David Allyn has pointed out, *The Joy of Sex* directed itself not at "young hippies" (who, in 1972, were generally thought to know all they needed to know about obtaining sexual pleasure) but at the older generation (Siegel). Even so, American high school students were still on the receiving end of sex instruction, although classroom films were becoming more graphic (see Eberwein for a discussion of early-1970s titles such as *Free* and *Achieving Sexual Maturity*), and Campbell observes that "authors [for the young] who cheerfully urged hip and carefree sex found themselves embarrassed and out of print very soon" (158).
8. See Campbell 87 and 115 for examples of American reprints in the 1920s and 1960s. Sally Lodge's April 2001 *Publishers Weekly* article on new editions of puberty texts from the 1970s and 1980s is one piece of evidence that the practice is still common; we thank Camille Phillips for providing this reference.

Works Cited

Ballard, L. Anna. "Danger to Our Girls." Social Purity Series Leaflets for Mothers' Meetings, No. 4. N.d. Rpt. in "L. Anna Ballard, M.D.: Pioneer Woman Doctor in Lansing, Michigan." Ed. D. La Pierre Ballard. 16 July 2001. Accessed 16 October 2001, http://balcro.com/anna.html.

Campbell, Patty. *Sex Guides: Books and Films about Sexuality for Young Adults.* New York: Garland, 1986.

Eberwein, Robert. *Sex Ed: Film, Video, and the Framework of Desire.* New Brunswick: Rutgers University Press, 1999.

Ewart, Gavin. "The Old School." *Out of Bounds: Public Schools' Journal Against Fascism, Militarism and Reaction* 1.2 (June–July 1934): 12–13.

Lodge, Sally. "Growing-Up Books Come of Age: Three Evergreen Guides to Maneuvering through Puberty Are as Popular as Ever." *Publishers Weekly* (2 April 2001): 23–26.

Moran, Jeffrey P. *Teaching Sex: The Shaping of Adolescence in the 20th Century.* Cambridge: Harvard University Press, 2000.

Morgan, Sue. "Faith, Sex and Purity: The Religio-Feminist Theory of Ellice Hopkins." *Women's History Review* 9.1 (2000): 13–34.

Nelson, Claudia. "Playing Patients: The Medico-Moral Context of *The Heavenly Twins*." *Journal of Contemporary Thought* (Summer 1999): 145–68.

——. " 'Under the Guidance of a Wise Mother': British Sex Education at the *Fin de Siècle*." *Maternal Instincts: Visions of Motherhood and Sexuality in Britain, 1875–1925.* Ed. Claudia Nelson and Ann Sumner Holmes. Houndmills, Basingstoke: Macmillan, 1997. 98–121.

Porter, Roy, and Lesley Hall. *The Facts of Life: The Creation of Sexual Knowledge in Britain, 1650–1950.* New Haven: Yale University Press, 1995.

Romilly, Giles. "Morning Glory (Sex in Public Schools)." *Out of Bounds: Public Schools' Journal Against Fascism, Militarism and Reaction* 1.2 (June–July 1934): 32–34.

Schemo, Diana Jean. "Surgeon General Calls for Sex Education Beyond Abstinence Courses." *New York Times* 29 June 2001, http://www.nytimes.com/2001/06/29/national/29SEX.html.

Siegel, David. Interview with David Allyn aired 28 March 2000. *All Things Considered.* Accessed 5 January 2002, http://www.npr.org/programs/atc/books_music/2000/mar.

[Turner, Graeme.] "Australian Cinema." Lecture notes posted July–September 2001. Accessed 5 January 2002, http://english.uq.edu.au/courses/ccst.2310/lecture1.htm and ff.

Van de Velde, Th. H. *Ideal Marriage: Its Physiology and Technique.* 1926. Trans. Stella Browne. New York: Random House, 1930.

Warne, Ellen. "Sex Education Debates and the Modest Mother in Australia, 1890s to the 1930s." *Women's History Review* 8.2 (1999): 311–27.

Wyndham, Diana H. "Striving for National Fitness: Eugenics in Australia, 1910s to 1930s." Ph.D. thesis. Sydney: University of Sydney, 1996. Accessed 4 January 2002, http://setis.library.usyd.edu.au/~thesis/adt-NU/uploads/approved/adt-NU2000.0015/public/whole.pdf.

Chapter 1

That Other Eden: Adult Education and Youthful Sexuality in *The Pearl*, 1879–1880

Claudia Nelson

Victorian sex-education manuals aimed at the young often sought to preserve the child's purity while still enlightening readers about where babies come from.[1] But as Roy Porter and Lesley Hall observe, sex-education manuals as conventionally defined were by no means the only vehicles for the transmission of sexual knowledge in nineteenth-century Britain: "At any moment, scores of books and pamphlets were available, offering direct or indirect sexual instruction, from encyclopaedic reference works to erotic prints and magazines" (5). Fraser Harrison, similarly, pairs pornography with medical tomes as the major forms through which Victorians promulgated such information (52). Both were aimed primarily at adult male audiences, and both were genres that circulated internationally; medical sexology in Britain owed much to Continental sources, while pornography was readily smuggled across the English Channel from Paris but also arrived from North America. The two forms might even emanate from the same press; for example, Charles Carrington, the Paris-based Victorian publisher of erotica, produced also more serious titles such as Dr. Benjamin Tarnowsky's *The Sexual Instinct and Its Morbid Manifestations from the Double Standpoint of Jurisprudence and Psychiatry* (1898).

Indeed, depending on the reader's response, the boundary between medical instruction and pornography was readily blurred. Recall the moment in chapter 20 of *The Adventures of Tom Sawyer* when Becky Thatcher peeps into her teacher's "mysterious" copy of "Professor somebody's 'Anatomy,' " about which the children's curiosity is at fever pitch, encounters a colored illustration of "a human figure, stark naked," and is so overcome by guilt and shame at the thought of discovery that she accidentally tears the picture (Twain 523). And if the medical tome had pornographic possibilities, pornography often foregrounded its pedagogical function, presenting itself as designed for the conversion or enlightenment of readers; consider, for instance, the subtitle of the 1882 novel *Mysteries of Venus,* which styled itself "A Preceptor for Ladies and Gentlemen on Their Wedding Eve" (Kincaid 181, n. 124). This attitude was not lost upon purity advocates. The Lord Chief Justice's 1868 definition of obscenity as matter exhibiting "the tendency . . . to deprave and corrupt" (qtd. Hynes 256) recognizes (and, of course, condemns) pornography's educational dimension: here, obscenity does not merely pander to immorality, but produces it, just as morally beneficial works inculcate virtue.[2]

In erotic magazines, the urge toward didacticism might take forms ranging from editorial direct address to self-referential plot twists such as one contained in the short story "How He Lost His Whiskers: An Episode in the Life of Steve Broad" (*Pearl* December 1880). In the latter work, a formerly censorious old man, inveigled into reading the latest number of the very periodical in which "How He Lost His Whiskers" appears because he hopes that the magazine will afford a clue to the identity of the sexual malefactor who has dropped it, declares that he has "quite altered [his] opinion about that young fellow [he] saw and [is now] convinced that to give his prick a treat is the first duty of every man at all times" (635). Still more typical is Victorian pornography's focus on the loss of virginity as an initiation into an enviable state of erotic fulfillment, the demarcation not between innocence and experience but between ignorance and bliss. If one American writer of 1894 emphasized his theme by giving the title *Maidenhead Stories* to his collection of young party guests' reminiscences about their sexual initiations, other works less overtly concerned with this issue nevertheless return to it again and again. Indeed, because the learning process itself is of obsessive interest, the nineteenth-century erotic text may not only be particularly dangerous for teens and preteens ("those whose minds are open to such immoral influences," as Lord Justice Cockburn put it in his decision in *R.* v *Hicklin* in 1868 [qtd. Hynes 256]), but also particularly interested in them. Such figures are represented as simultaneously childlike and sexual, all the more appetizing because of their untouched quality.

Thus while historian Peter Gay has argued that Victorian erotica "reads like a plagiarism of infantile fantasies" in its regressive focus on voyeurism, scatology, and father–daughter incest (373), texts such as *The Pearl* (1879–80), the underground magazine on which this chapter will focus, do not limit themselves to transplanting the sexual obsessions of childhood into adult settings.[3] More directly, they also present childhood desires as mature and compelling, and indeed as superior in many ways to conventional adult sexual expression. In detailing children's hetero- and homosexual liaisons with servants, racial Others, schoolmates, teachers, and family members, *The Pearl* seeks, of course, to titillate readers. But this titillation depends not merely upon the upending of post-Romantic stereotypes of childhood innocence and purity, or upon the ratification of pre-Victorian stereotypes of childhood indiscipline, primitivism, and amorality, but rather upon the blending of the two. If children are sometimes rendered as the sexual pupils of their elders, vulnerable to seduction, trickery, and predation, they are equally likely to appear as sexual guides. The qualities ascribed to them, notably the same lack of inhibition, eagerness for experience, and preternatural talent at the dissimulation made necessary by puritan sexual mores that Becky Thatcher exhibits in the episode from *Tom Sawyer* mentioned earlier, are presented as at once delicious and worthy of emulation by adults. *The Pearl*, then, may be read as a manifestation of sex education aimed principally at well-to-do adult males and depending on (and consequently exposing) contradictory cultural assumptions about what upper-class Victorian children "naturally" were.

The Pearl in Its Setting: Victorian Responses to Childhood

For respectable Victorians as for the demimonde, images of childhood were both complex and conflicted. On the one hand, we may discern a strong belief in the moral innocence of the middle-class child and in that child's resulting suitability as a model for adult behavior; on the other, we also see manifestations of the early nineteenth-century understanding of children as naturally unregenerate, desperately in need of guidance from their elders and betters. Popular tales such as Hesba Stretton's *Jessica's First Prayer* (1867), George Eliot's *Silas Marner* (1861), and Frances Hodgson Burnett's *Little Lord Fauntleroy* (1885) marketed the image of the redemptive child who, all unconscious, woos a solitary man away from selfishness and misanthropy. Simultaneously, writings on pedagogy, child rearing, and health warned readers that children, lacking a strong moral compass, are liable to corruption, so that adults must exercise constant vigilance to keep them on the right track. The questions of who should be the teacher, who the student, and what the curriculum are perennial in texts in which the generations meet.

This uncertainty is particularly evident in Victorian discussions of childhood sexuality. Depictions of the post-Romantic child often highlighted that child's supposed freedom from sexual impurity, an innocence implicitly presented as an object lesson for adults. Although working-class children were often assumed to be sexually precocious because of the opportunities afforded by their cramped quarters and unrefined surroundings, many observers deemed the offspring of the privileged "above" sexuality just as they were "above" other taints that Victorian social critics found within their culture, such as commercialism or religious doubt. Yet the child's asexuality could not be taken for granted. For instance, Steven Marcus notes that while William Acton, a medical writer heavily invested in the purity of children and women, opined in 1857 that "perfect freedom from, and indeed total ignorance of, any sexual affection is, as it should always be, the rule" in the young, Acton followed a brief discussion of childish innocence with a lengthy homily that "communicates unmistakably the sense that childhood sexual play and childhood masturbation were both widespread and well-known phenomena" (13–14, 15).

Acton differentiated the "normal" asexual child from the "abnormal" sexual child. Sex experts of subsequent generations, however, did not always draw this distinction, suggesting rather that sexual expression and childlike innocence might coexist. Richard von Krafft-Ebing remarked in 1879 that "manifestations of sexual instinct may occur [even] in very young children" but should be considered "premature" only when they are "a result of cerebral processes without peripheral causes" (36); similarly, shortly after the turn of the century, Havelock Ellis invoked earlier authorities to comment that "rudimentary sexual activities in childhood, accompanied by sexual feelings, must . . . be regarded as coming within the normal sphere" and should be considered "play" compatible with "genuine innocence" (II.3.36–37).

Such a juxtaposition of eroticism and purity might serve, again, as a potential model for middle-class adult behavior. Historian Eric Trudgill has noted that for many Victorians—Lewis Carroll and John Ruskin are the obvious examples, but they were far from unique—the very innocence of the "child angel" meant that she, or sometimes he, could provide "a degree of subdued sexual release" (94). Men seem often to have been attracted to children not out of a perverse desire to defile, but rather out of a longing to reenter a lost Eden. Conversely, women might be encouraged to cling to childhood not only because their early training left them poorly equipped for the complexities of the grown-up world, but also because it enhanced their marketability; Charles Dickens's "sexy but infantile" Dora in *David Copperfield* (1850) is one of many mid-Victorian creations to combine the role of child with the role of erotic object (Harrison 38–45). As Carroll observed to the artist who was to illustrate *Sylvie and Bruno* (1889), "naked children are so perfectly pure and lovely" (qtd. Trudgill 93). It was this perceived purity that formed the basis of the erotic appeal of the prepubescent child in Victorian England—and, indeed, at other times and places, as Shurlee Swain, Ellen Warne, and Margot Hillel's discussion of the 1918 Australian children's novel *St. Tom and the Dragon* makes clear (see chapter 2 in this volume). Carroll's photographs of naked children celebrate purity, in that he displays his models as unselfconscious and unhampered by the taboos or inhibitions or shibboleths associated with acculturation into the adult world, but they simultaneously celebrate sexuality as well.

Today's viewers often classify Carroll's child nudes as works that are colored by pedophilia without being pornographic. Yet that Carroll gave up photographing nudes in 1879 and destroyed most of the nudes that he had taken earlier acknowledges the probability that the images would be perceived as sexual even (or especially?) within a culture that insisted on the child's purity.[4] And indeed, in their twin focus on purity and sensuality, the photographs share a characteristic concern of the pornography of their day. Frank Mort, for instance, notes that in "photographic studies of child prostitutes dating from the 1860s and 1870s . . . childhood innocence was erotically framed against visible signs of immoral sexuality, such as exposed genitalia or the depraved stare" (84). In some contexts, purity might appear not as availability's opposite but as its preamble.

According to Marcus, "Pornography exists in order to violate in fantasy that which has been tabooed" (245); if we accept this dictum, we might choose to see the pedophiliac tendencies of *The Pearl* as a reaction against a society that placed childhood "above" sexuality. But as I have already suggested, this reading may be too simple to meet the case. Just as the morally uplifting figure of the Angel in the House necessarily possessed a sexual side because the Angel's ultimate function was to bear and raise children (and James Eli Adams, drawing on the work of Michael Mason, notes the "widespread belief that women must experience orgasm to conceive" [131]), the figure of the redemptive child could be regarded in sexual terms as part of a larger Victorian project of reconciling sensuality and virtue. Eroticism, after all, was respectable as long as it remained vague, so that passionate friendships between schoolboys were deemed uplifting, and sensuous paintings of young children were deemed beautiful.

Conversely, nineteenth-century society arguably considered children less taboo sexually than is the case today. Swain, Warne, and Hillel, for instance, note in the present volume (chapter 2) that Australians charged with rape and sexual abuse in the 1890s might convince courts of law that girls in their early teens were "consenting adults." James Kincaid proposes that perhaps "the Victorians, busy sexualizing the child, still did not invest in the process the degree of anxiety common in [the twentieth] century" (201). And, along similar lines, Ronald Hyam suggests that the sizable population of juvenile prostitutes, visible to observers long before W. T. Stead and other reformers targeted it for international attention in the 1880s, was larger in Britain than elsewhere in the world (62). Until 1875, the age of consent for girls under English law—this milestone did not exist for boys—was twelve; between 1875 and 1885, it was thirteen. If in *John Marchmont's Legacy* (1863) Mary Elizabeth Braddon can emphasize her heroine Mary's childishness by perching her, as a "little girl" of thirteen, on her father's knee (50), Braddon sees nothing paradoxical in earlier establishing that at age eight, Mary's "womanly heart" is capable of forming a lasting passion for her future husband (40). Victorians of all levels of respectability and throughout the English-speaking world were alive to the appeal of what Braddon terms the "child-woman" (47), an ideal that insisted on the naturalness of romantic attachment in an era in which love was elsewhere codified, formalized, and generally rendered as "unnatural" as possible.

Child's Play: *The Pearl's* Construction of Sexuality

If in one sense the aim of *The Pearl* is to shock (and thus to please) the bourgeois, then, in another it is to urge the rejection of "adult" sexual codes: those that have to do with proscription, to be sure, but also those that define sexuality as inextricable from procreation. A scene in the June 1880 installment of "Lady Pokingham, or They All Do It" is emblematic both in its argument and in its representation of pornography as pedagogy. Recently widowed by the death of Lord Crim-Con, the heroine seduces her brother-in-law, Robert, determined to turn "the hitherto respectable husband . . . into a lustful libertine" because she dislikes his wife (394). Robert, now Lord Crim-Con by inclination as well as inheritance, welcomes his wife home with the news that "during your absence I found some bad books of my late brother's and they so fired my imagination by the extraordinary descriptions of various modes of sexual enjoyments that I quite blushed to think of our innocent ignorance, and long to try some of them with you" (423–24).[5] His motivation, he explains, is not so much the craving for new sensation as it is the wish to experience sensation without consequences, since "our plain silly way of doing it only leads to getting a lot of children, and surely my quiver is full enough of them. I'll have no more, it's positive ruination, however rich a father may be" (424). On the one hand, in other words, the sexual education acquired by reading pornography (or having an affair with the heroine of an erotic serial) provides one with a wider range of options, not only in bed but also financially. As a number of historians have observed, the double meaning inherent in the Victorian use of the verb "to spend" reflects a persistent nineteenth-century conflation (see, e.g., Adams 127).[6]

On the other hand, that this knowledge provides access to forms of sexual expression that do not involve parenthood simultaneously implies a return to a childlike state in which all sex is only "play," to recall Ellis's term. It is perhaps suggestive that while avoiding pregnancy is a recurring theme, no artificial aids to contraception (too "adult"?) appear in any contributions to *The Pearl*, although "My Grandmother's Tale or May's Account of Her Introduction to the Art of Love" contains one reference to douching (412) and several to *coitus interruptus*.[7] And here again the pedagogical aspect of the magazine is highlighted, since some of these references are highly didactic in inclination. Consider the following passage from April 1880:

> Then he explained how that [*sic*] unless the seed was injected on the mouth of the womb, which lay at the end of the passage, there was no danger of any woman being put in the family way. And though the pleasure of both parties is lessened by the withdrawal of the prick at the moment of highest enjoyment, yet a man must be a selfish brute if on that account he would run the risk of doing such a grievous wrong to any girl whom he respected and loved. (384)

The fatherly tone of these remarks seems directed more at the neophyte male reader than at the ostensible audience, the two girls in their mid-teens with whom the adult speaker is occupied; at issue here is both the formation of a particular attitude in heterosexual men and the imparting of information that will permit the latter to indulge themselves without fear. The sexual subject, we are assured, need not be thrust into the role of paterfamilias or angelic mother, but may stay young and carefree in a sexual Neverland.

Accordingly, what is striking about the protagonists of the various stories and serials contained in *The Pearl* is not that some are young but that nearly *all* are. In this emphasis on youth, *The Pearl* resembles rival works such as *The Romance of Lust* (1873–76), *The Romance of Chastisement* (1866), and *The Amatory Experiences of a Surgeon* (1881) by "James Campbell." In the last-named work, which shared a publisher with *The Pearl*, the hero, the love-child of a titled aristocrat, begins his sexual life at his public school at age fourteen and, having embarked upon a medical career, "is invariably overcome by lust when he examines barely pubescent female patients," some of these " 'unripe beaut[ies]' " being as young as thirteen (Anderson 99, Marcus 235–37). Like all these narratives, *The Pearl* makes much of the erotic possibilities of the two great educational environments available to well-to-do children, the family and the boarding school.[8] This focus, and the concomitant oscillation between academic discipline (in the sexual sense of the term) and domestic coziness, may help to underscore the magazine's stated wish to position itself as a pedagogical text: that characters constantly move back and forth between the role of student and the role of teacher implies to readers not only that both dominant and subordinate positions have their erotic advantages, but also that the didacticism so inescapable in Victorian culture may be read, perversely, as masking sexual impulses.

Thus, with the exception of a single short story featuring a middle-aged sultan and a thirty-year-old sultana valide, every *Pearl* prose selection above anecdote length features at least one major character aged twenty or below.[9] Of the serials, for instance, "Sub-umbra, or Sport Among the She-Noodles" follows the exploits of a family group of boys and girls ranging in age from fifteen to nineteen; the title character of "Miss Coote's Confession" begins her reminiscences by describing her domestic and academic experiences at age twelve; Lady Pokingham's first sexual encounter likewise occurs when she is twelve and involves a schoolmate who lost her virginity when, at ten, she seduced the family butler; the hero of "La Rose d'Amour" is seventeen when we meet him; "My Grandmother's Tale" concerns two sixteen-year-old pupil-teachers and a nineteen-year-old named Kate, who was initiated nine years previously by a boy of her own age; and Charles, the narrator of "Flunkeyania; or Belgravian Morals," is doubly unique in being the only person as old as twenty and the only servant to function as protagonist of a *Pearl* serial. Shorter narratives contain children as young as nine and ten. And while the ten-year-old appears as a victim of sexual assault in what purports to be a court transcript ("The Trial of Captain Powell," January 1880), nine-year-old Janey in "Young Beginners" (October 1879) first happily watches and subsequently participates in explicit and repeated sex play with her brothers and a cousin.

Today's readers, of course, will discern a significant difference between the sexuality of a nine-year-old and that of a nineteen-year-old. The one is unequivocally a child by our standards, the other a young adult—potentially, indeed, a parent. But while Victorian observers too would have conceded that someone nearly twenty is at least virtually an adult, it is important to note that ways of constructing age have changed over time. The term "adolescent," popularized in the twentieth century, is generally used today to describe any individual between the ages of, say, thirteen and eighteen, but originally denoted "the years from sexual maturity to the end of physical growth in a person's twenties" (Weeks 50). In the nineteenth century, the duration of childhood had much more to do with social class and gender than is now the case; a twelve-year-old servant was for all intents and purposes an adult, while a twelve-year-old aristocrat was not. Upper-middle-class boys, who were likely to leave home for boarding school at age seven, were often presumed to be more sexually sophisticated than their twin sisters educated at home; a "home daughter" might well linger in childhood's anteroom, doing lessons in a desultory fashion and subscribing to girls' magazines, until her early twenties.

Indeed, as Jeffrey Weeks observes, dietary considerations dictated that physical maturity tended to come later in the nineteenth century than it does today: "The ages of puberty and menarche were at least as high as 16 for the rural population in the early nineteenth century" (48). While by 1879 the well-nourished children of the comfortable classes would have reached puberty earlier, social expectations would nonetheless have worked to emphasize the youth of such individuals.[10] Significantly, in the 1870s members of the upper middle class married later than people in other walks of life (Weeks 73); while this trend resulted partly from economic considerations (setting up one's own establishment was far more expensive than bachelorhood), society could encourage it by defining sexuality and youth as incompatible. It thus should not surprise us that the narratives contained in *The Pearl* describe fifteen-year-olds such as Polly in "Sub-umbra" as lacking pubic hair. Both physically and socially, Polly is officially a child—but, despite her class membership, a child who is sexually active. In eschewing working-class protagonists, *The Pearl* is effectively identifying its teenage characters as operating outside adulthood on every level but the sexual.

Nor is the sexual maturity demonstrated by Polly and her ilk a result of contamination by those more precocious social inferiors. Ellis mentions "the great part played by servant-girls of the lower class in the sexual initiation of the children of the middle class" (II.3.43); in the present volume (chapter 2), Swain, Warne, and Hillel quote the fears of an Australian commentator of 1918 that servants would provide sexual information to their employers' children. Both moments echo a fear common in Victorian England: that the domestic purity of the respectable household would be sullied by the introduction of representatives from the lower orders, often nursemaids who would stimulate their charges while the latter were still in leading-strings. But another striking circumstance in the narratives contained in *The Pearl* is that here the seduction invariably goes in the other direction; while servants may be attracted to their social betters, respect keeps them in their place until the well-born extend an unmistakable invitation. Alice Marchmont in "Lady Pokingham" must extort sexual favors from the butler by threatening to reveal his liaison with another servant. Lady Pokingham herself acquires a penchant for servants at age seventeen, during an orgy in which aristocratic women make use of teenage employees, and later sets herself the challenge of taking the virginities of the six members of her staff, most of whom are fifteen or sixteen years old. Rosa Coote, at twelve one of the objects of her grandfather's flagellation, soon whips and afterward beds a female servant who is represented as a comparatively passive partner in this conjunction. Conversely, in "A Black Joseph" (November 1879), one of the historical courtroom accounts that dot the pages of *The Pearl*, we hear that "our youthful Negro [footman] of nineteen manfully withstood" the attempts of his employer's wife to make use of him sexually (169).

The effect of such episodes is not only to tweak audience expectations by inverting certain clichéd domestic anxieties of the era, or to note the erotic possibilities of class differences and the virility of the peasantry, but also to establish unfettered sexual expression as a privilege of the upper classes. To associate seduction with "servant-girls of the lower classes" is to stamp it vulgar and degrading, an activity that the civilized have risen above. To link it, as happens in *The Pearl*, with wealth and power (although not necessarily with masculinity or maturity) makes it an ideal to be emulated. Tellingly, sexual encounters in these narratives are typically consensual and based on mutual desire; certainly rape is by no means unusual—although rape victims, predictably, rapidly turn into willing participants—but it is not the dominant mode here, and prostitution is nonexistent. Moreover, the pervasive Victorian fear that sex is enervating and destructive to the individual (as to the corporate) body is significant only in "Lady Pokingham," in which both the title character and her husband meet untimely deaths because of their addiction to venereal pleasure. Syphilis and gonorrhea are nonexistent, and orgasm is often represented in downright medicinal terms as "most invigorating" (385) or "a warm and soothing enema" (410).[11] In a prefiguration of the early twentieth-century pro-pleasure marriage manuals written by upper-middle-class professional women and described by Hall in the present volume (chapter 3), the effect is to suggest that this is how the Beautiful People live, and that readers—and their wives, children, and servants—should go and do likewise. The serials, in short, construct an idealized England in which sex is free, readily available, chic, and without undesirable consequences.

Getting Away with It: The Pedagogy of Dissembling

To be sure, *The Pearl* simultaneously acknowledges that the enthusiastic and guilt-free couplings it chronicles take place against a backdrop of social disapprobation, and this circumstance, too, helps to shape the magazine's "educational" message and to dictate its preference for youthful protagonists. For the sexual actors here are aware that gratification may require dissimulation and a degree of rebellion. Like the children in a number of more respectable Victorian narratives, the young people of *The Pearl* are free of the trammels of convention, and happier for it. Thus they model the tutelage given in the editor's preface to *The Pearl*'s inaugural number: readers "have only to keep up appearances by regularly attending church, giving to charities, and always appearing deeply interested in moral philanthropy, to ensure a respectable and highly moral character, and . . . if they only are clever enough *never to be found out*, they may, *sub rosa*, study and enjoy the *philosophy of life* till the end of their days" (1, emphasis in original).

As Marcus notes, Victorian pornography not infrequently implies "that hypocrisy is a very good and very useful thing indeed," so that in a sense it functions to expose "the discrepancy which exists in society between openly professed ideals and secretly harbored wishes or secretly practised vices—it may act indirectly to 'unmask' society's official version of itself" (229, 230). In this regard the genre resembles the sensation novels of the 1860s, which likewise, as Toru Sasaki and Norman Page note in their introduction to *John Marchmont's Legacy*, focus on "crime within the family, secrets behind the façade of middle-class respectability" (vii).[12] But in *The Pearl*, this unmasking seems intended not to chastise hypocrisy but to encourage it by suggesting that such behavior is common and justifiable; part of the pedagogical function that the magazine claims for itself is to teach methods of getting away with naughtiness.

Indeed, one such method is simply to occupy a role associated with virtue, such as those of child and/or virgin. Ten-year-old Alice Marchmont, for instance, is assisted in her pursuit of the butler by his assumption that "Alice could know nothing." Because he sees her as a sexual innocent, he has no idea that she may want to put an end to her own ignorance until she finally corners him in a boat and asserts her right to enlightenment. In a passage suggesting that the best sex education has an empirical component, Alice is shown insisting, "You shall tell me what I want to know this time: How do babies come? What is the parsley bed, the nurses and doctors say they come out of? Is it not a curly lot of hair at the bottom of the woman's belly? I know that's what Lucy's got, and I've seen you kiss it, sir!" (26–27). Other girls encourage male sexual aggression by feigning sleep or unconsciousness, states of ostensible purity as complete as they are temporary. While *Pearl* serials frequently refer to virginity as "troublesome," a burden to be shed as quickly as possible, in another sense it is clearly helpful to its possessors in that it enables them to attract partners—not only because the "spendings" of virgins both male and female are described as unusually luscious, but also because virginity is represented as a state in which lust and purity coexist. Inherently hypocritical, virginity is therefore peculiarly sexual.

Along similar lines, we witness the central figures of "Sub-umbra" manufacturing opportunities for sexual encounters not only by direct methods such as drugging their parents (184), but also by participating in "polite" activities that they then turn to their own ends, such as strolling in the grounds or singing genteel drawing-room ballads (108). Stereotypes of youthful innocence and the innocuousness of the pastimes of the well-bred young are the characters' friends here, so that the story (and, indeed, the magazine as a whole) teaches that assumptions about childhood innocence are precisely what enable less-than-innocent behavior. Even so, and perhaps because the narrator of "Sub-umbra" is male, this serial suggests a frontal assault upon the authority of the previous generation; stories with female protagonists are more likely to depict a conspiracy between girl and older man to flout the more abstract authority of society.

For instance, "Kate's Narrative," a segment of "My Grandmother's Tale," deals with father–daughter incest by showing Kate, just entering puberty, impersonating her father's fifteen-year-old slave mistress, Nina, in order to receive his sexual addresses. When he taxes her with the deception the following day, she confesses and is readily forgiven:

> "Tell me, my darling, was it you in Nina's bed last night?"
> "It was I, dearest papa; was I very wicked?"
> "No, my darling, you gave me the sweetest pleasure I ever had in my life." (449)

The two then bed one another again, Nina joining them along the way. As the phrases "dearest papa" and "sweetest pleasure" imply, Kate is to be construed as a dutiful daughter, indeed a virtuous daughter, who has not tricked her father into incest so much as she has relieved him of the burden of decision in their mutual subversion of social mores. That the decision is

Kate's would nonetheless seem to establish her as the moral authority within the saga. Youthful lack of inhibition is necessary if one is to profit from adult expertise.

The Question of Gender

The summaries that I have provided of the experiences of Kate, Lady Pokingham, Alice Marchmont, Rosa Coote, and so on should already have made evident that *The Pearl* frequently depicts female characters—even, sometimes, prepubescent virgin girls—as the initiators in sex, as adept as their male counterparts in the scheming and subterfuge necessary to construct erotic opportunities. Indeed, as Patricia Anderson has pointed out, the original subscribers would have understood the magazine's name as a synonym for the clitoris (8), so that female pleasure was foregrounded on every cover. Depictions of lesbian activities in serials such as "Lady Pokingham" represent the organ as essentially masculine, a way for phallic delights to be accessible to women as well as to men, as in the following passages from the December 1879 number: "my touches . . . had caused quite an erection of her finely developed clitoris"; "I entreated her to be the gentleman, and let me have her stiff clitoris, which I was sure could give me great pleasure"; "[she had a] splendid mount covered with glossy black hair . . . from which projected quite four inches a stiff fleshy clitoris as big as a man's thumb" (199). Lawrence Birken has observed that for late-nineteenth-century sexologists, "the clitoris was linked to the ancestral hermaphrodite, [so that] it was something the truly 'feminine,' and thus truly differentiated, woman was expected to 'overcome' " (82). In contrast, *The Pearl* consistently approves women's "hermaphrodite" qualities, rejecting cultural prescriptions about female passivity in favor of the insight that, in Ellis's words, the "girl [is] . . . often the one who has taken the more active part in initiating the act" (II.3.529).

Nevertheless, it is difficult to take this periodical as preaching female sexual fulfillment in any uncomplicated way, á la some of the sexual radicals of the 1890s and earlier, since, like the vast majority of Victorian erotica, it presumably aimed at a readership overwhelmingly male. (Significantly, all the characters represented in *The Pearl* as possessing works of pornography are adult men.) The serials do not seem designed to establish that women are more likely than men—or girls more likely than boys—to be sexually aggressive. Rather, the pattern is merely that any narrator, whether male or female, adult or child, will take an active role. And presumably as an aid to reader identification, all the stories above anecdote length are told in the first person.

In other words, male readers were encouraged to imagine themselves not only as supremely well-endowed and virile young men, but also as lesbians, flagellatrixes, nymphomaniacs, and seductive small girls, depending upon the identity of the central figure of the particular narrative in question. Nevertheless, they were still to be socialized as Victorian convention expected: as "instrumental, forceful and direct" (Weeks 42), the dominant partners in sex. Marcus has written of nineteenth-century flagellation tales that as a genre, they are characterized by "childishness, extreme incoherence, absence of focus, confusion of sexual identity, and [an] impulse toward play-acting or role playing," traits that could also be ascribed to *The Pearl* overall, and that "these qualities stand in marked contrast to the Victorian ideals of manliness, solidity, certitude of self, straightforwardness, sincerity and singleness of being" (263). While this insight is valuable, it is important to add that what Victorian science often construed as the central traits of masculinity, namely activity and energy, are still valorized for male readers; maleness was not to undergo the ultimate deconstruction. In some regards, at least, the sex education imparted by *The Pearl* ran along conventional lines.

Conclusion: The Pedagogy of Difference

The Pearl's obsessive focus on the joys of defloration insists upon the link between instruction and delight. At least where sexuality is concerned, the magazine tells its readers, both learning and teaching are sources of ecstasy. Less direct than its emphasis on the education of the characters who populate its fictions, the education that *The Pearl* implicitly offers to its audience nonetheless also promises pleasure, however illusory or vicarious. And the source of this pleasure is to be the mingling of sameness and difference whose primary locus, in these pages, is the child. In identifying with the boys and girls whose sexual initiations are narrated in the various serials, adult male readers are confirmed in their understanding that sexuality should fit the authoritarian model for masculine reproductive behavior identified by Ellis and earlier writers such as biologists Patrick Geddes and J. Arthur Thomson. At the same time, that readers are encouraged to identify with figures who are, at least ostensibly, *not* adult males suggests a rejection of particular adult male functions.

The most obvious of these functions is the power to impregnate. Perhaps more closely allied to the pleasure principle, however, is these narratives' focus on the erotic potential latent in the lack of social power—in being, as well as in possessing, the Other. We may find it easy to determine who is powerful and who powerless in an encounter between a full-grown middle-class man and his young female servant, say. But while such encounters certainly occur in the pages of *The Pearl*, they generally take place in the background, helping to establish a general atmosphere of lubricity rather than providing the principal erotic focus of a chapter. Much more interesting to *Pearl* contributors, and thus demanding more attention from readers, are those moments in which the apportioning of power is confused. Where does authority lie when a sexually experienced adult male servant is coerced into taking the virginity of a ten-year-old aristocratic girl for whom he has always entertained warm feelings? When a young Black manservant successfully refuses the advances of his employer's wife? When a schoolgirl birches her racially mixed teacher in a sex game before agreeing to the teacher's importunities? When a wealthy woman is so entranced by the sight of a teenage brother and sister, her employees, in an incestuous embrace that she determines to abandon her erstwhile predilection for female partners, first flogging the boy and then crying out in passion, "Ah! the punishing of you has been the undoing of me, ah! I am indeed afraid of you" (347)?

The answer, apparently, is that authority is vested less in the participants than in sexuality itself, routinely characterized as an irresistible force far more powerful than masculinity, adulthood, or wealth. In order to establish this point, *The Pearl* finds it necessary to explore multiple varieties of difference, including differences of class, race, gender, and age. But it is the age factor that seems to trump all the others. Youth is the one constant in these tales, and the traits ascribed to it, most importantly indefatigable sexual appetite combined with the willingness to defy convention and the ability to retain the appearance of innocence while doing so, are held up as copybook models for the improvement of readers of mature years. This enterprise is the more convincing because it extends and parodies, but does not contradict, maxims about childhood prevalent within the larger culture. In their self-proclaimed effort to educate an adult readership, Victorian pornographic works such as *The Pearl* relied upon lessons that respectable as well as dissolute audiences had already learned.

Notes

1. To be sure, those aimed at pubescent schoolboys often took a less idealistic tack, assuming that the boy's purity had already been compromised by the exigencies of public-school life and engaging in warnings or threats that imply, in Steven Marcus's words, that far from trailing clouds of glory, children are "in danger of becoming little monsters of appetite" (15). For a discussion of these two schools of sex manual, see Nelson.

2. This understanding of obscenity was what enabled the prosecution and potential banning of texts intended more to teach than to titillate, including birth-control manuals, sexological studies such as the work of Havelock Ellis, and (after the passage of the Indecent Advertisements Act of 1889) "any advertisement relating to syphilis, gonorrhea, nervous debility, or other complaint or infirmity arising from or relating to sexual intercourse" (qtd. Hynes 256). As Samuel Hynes points out, the original Victorian legislation permitting the suppression of pornography—the Obscene Publications Act of 1857—furnished no definition of what pornography might be, but rather left this question to the judgment of the individual magistrate (255).
3. The Victorian publication history of *The Pearl* remains obscure; little is known about circulation figures, contributors, or original readers, and Patrick J. Kearney christens its publisher—who also issued novels such as *Randiana, or Excitable Tales* (1884)—"the mysterious Lazenby." It is easier to track the magazine's post-Victorian life: not only will Internet searches turn up a number of tributes by recent writers of erotica who credit *The Pearl* with influencing their own work, but we also find that mainstream publishing houses in both Britain and the United States have republished the periodical, wholly or in part, in volume form. The edition that I am using, whose 643 pages contain the entire contents of the magazine (including the rhymes and jokes sometimes excised from modern reprints as insufficiently arousing), was published under the aegis of the Book-of-the-Month Club.
4. Four such images by Carroll survive. For a discussion of his photograph (ca. 1878) of Evelyn Hatch, then aged about seven and posed stretched across a bed of grass with her head pillowed in her hands and her body suggestively arched, see Mavor 159–70.
5. The nobleman's title refers to the legal abbreviation for "criminal conversation," or adultery. The books that the new Lord Crim-Con mentions are French, as is the supersexed hero of another *Pearl* serial, but at least one series of pornographic prints cited elsewhere in the magazine—"a set of scenes between a handsome white girl and a negro" (385)—evokes the Americas. The effect of these and similar devices is to represent sexual sophistication as simultaneously illicit and international.
6. In this context, it is interesting that the curmudgeons reformed by contact with children such as Eppie Marner, Lord Fauntleroy, and the eponymous heroine of *Jessica's First Prayer* are all, to some extent, miserly. By developing emotional bonds with the attractive youngsters, each man learns to "spend" in a morally appropriate fashion and thus to become better integrated into society.
7. Marcus notes that some Victorian pornography—he cites *Rosa Fielding, or, A Victim of Lust*, an 1867 novel published by William Dugdale and sometimes attributed to journalist G. A. Sala—does include advice on birth-control devices (234). Conversely, although Michael Mason reports that "married and unmarried couples at various levels of society adopted artificial contraceptive methods as early as the 1860s and these practices spread widely and quickly" (7), birth-control manuals such as Charles Knowlton's *Fruits of Philosophy* (U.S. 1832; U.K. 1834, rpt. 1877) and H. A. Allbutt's *Wife's Handbook* (1886) were sometimes prosecuted under the Obscene Publications Act. The divorcing of intercourse from procreation was no light matter.

8. Although in reality, boarding schools were much more a feature of boys' lives than of girls', *The Pearl* is particularly interested in female academies, presumably partly because lesbian adventures would have seemed more exotic and thus more titillating to its original audience than schoolboy homosexuality.
9. *The Pearl*'s "facetiae" consist of bawdy verse, jokes, acrostics, and so on, which often do not share the focus on youthfulness present in the rest of the magazine but which also do not seem intended as erotically compelling. They form part of a tradition of jestbooks on sexual themes, extending back centuries but in decline by 1879.
10. Although the House of Lords was reluctant to raise the age of consent to thirteen (members clearly assumed that sexually active girls were more likely to be their servants than their daughters), social-purity advocates—W. E. Gladstone among them—wanted it set at eighteen. The latter position is emblematic both of the effort to establish middle-class mores as the norm and of the belief that girls below eighteen should be classified as children (see, e.g., Hyam 65–66, Ellis II.3.528).
11. Mort notes that similar attitudes could be found in some medical handbooks; for Dr. Michael Ryan in the 1830s, for instance, "a moderate exercise of the sexual function for both marriage partners was essential to good health as well as conjugal happiness regulating 'the functions of the brain or nervous system and consequently of the whole body.' He condemned the practice of marital chastity as highly dangerous, leading to infertility in women, impotence in men and even to death" (80).
12. The resemblance seems also to have struck Victorian critics of the sensation novel such as Henry Mansel, future dean of St. Paul's, who in an 1863 *Quarterly Review* article charged thrillers in sexualized terms with being "indications of a wide-spread corruption, of which they are in part both the effect and the cause; called into existence to supply the cravings of a diseased appetite, and contributing themselves to foster the disease" (qtd. Sasaki and Page viii).

Works Cited

Adams, James Eli. "Victorian Sexualities." *A Companion to Victorian Literature and Culture.* Ed. Herbert F. Tucker. Malden, MA: Blackwell, 1999. 125–38.

Anderson, Patricia. *When Passion Reigned: Sex and the Victorians.* New York: Basic, 1995.

Birken, Lawrence. *Consuming Desire: Sexual Science and the Emergence of a Culture of Abundance, 1871–1914.* Ithaca: Cornell University Press, 1988.

Braddon, Mary Elizabeth. *John Marchmont's Legacy.* 1863. Oxford: Oxford University Press, 1999.

Ellis, Havelock. *Studies in the Psychology of Sex.* 2 vols. 1906 ff. New York: Random House, 1936.

Gay, Peter. *The Bourgeois Experience: Victoria to Freud.* Vol. 1, *Education of the Senses.* New York: Oxford University Press, 1984.

Hall, Lesley. "Eyes Tightly Shut, Lying Rigidly Still, and Thinking of England? British Women and Sex from Marie Stopes to Hite 2000." Included in the present volume, chapter 3.

Harrison, Fraser. *The Dark Angel: Aspects of Victorian Sexuality.* London: Sheldon, 1977.
Hyam, Ronald. *Empire and Sexuality: The British Experience.* Manchester: Manchester University Press, 1991.
Hynes, Samuel. *The Edwardian Turn of Mind.* 1968. Princeton: Princeton University Press, 1975.
Kearney, Patrick J. "Brandon House Library Editions: A Provisional Checklist." 2000. Accessed 29 January 2003, http://www.sonic.net/~path/Brandon3.htm.
Kincaid, James R. *Child-Loving: The Erotic Child and Victorian Culture.* New York: Routledge, 1992.
Krafft-Ebing, Richard von. *Psychopathia Sexualis: With Especial Reference to the Antipathic Sexual Instinct.* 12th ed. Trans. Franklin S. Klaf. New York: Bell, 1965.
Marcus, Steven. *The Other Victorians: A Study of Sexuality and Pornography in Mid-Nineteenth-Century England.* New York: Basic Books, 1966.
Mason, Michael. *The Making of Victorian Sexuality.* New York: Oxford University Press, 1994.
Mavor, Carol. "Dream-Rushes: Lewis Carroll's Photographs of the Little Girl." *The Girl's Own: Cultural Histories of the Anglo-American Girl, 1830–1915.* Ed. Claudia Nelson and Lynne Vallone. Athens: University of Georgia Press, 1994. 156–93.
Mort, Frank. *Dangerous Sexualities: Medico-Moral Politics in England Since 1830.* New York: Routledge, 1987.
Nelson, Claudia. " 'Under the Guidance of a Wise Mother': British Sex Education at the *Fin de Siècle.*" *Maternal Instincts: Visions of Motherhood and Sexuality in Britain, 1875–1925.* Ed. Claudia Nelson and Ann Sumner Holmes. Basingstoke: Macmillan, 1997. 98–121.
The Pearl: A Journal of Facetiae and Voluptuous Reading. 1879–80. New York: Book-of-the-Month Club, 1996.
Porter, Roy, and Lesley Hall. *The Facts of Life: The Creation of Sexual Knowledge in Britain, 1650–1950.* New Haven: Yale University Press, 1995.
Sasaki, Toru, and Norman Page. Introduction. Braddon, *John Marchmont's Legacy.* vii–xxiv.
Swain, Shurlee, Ellen Warne, and Margot Hillel. "Ignorance Is Not Innocence: Sex Education in Australia, 1890–1939." Included in the present volume, chapter 2.
Trudgill, Eric. *Madonnas and Magdalens: The Origins and Development of Victorian Sexual Attitudes.* London: Heinemann, 1976.
Twain, Mark (Samuel Langhorne Clemens). *The Adventures of Tom Sawyer.* 1876. Rpt. *The Unabridged Mark Twain,* Vol. 1. Ed. Lawrence Teacher. Philadelphia: Running Press, 1976. 436–585.
Weeks, Jeffrey. *Sex, Politics and Society: The Regulation of Sexuality Since 1800.* 2nd ed. New York: Longman, 1989.

Chapter 2

Ignorance Is Not Innocence: Sex Education in Australia, 1890–1939

Shurlee Swain, Ellen Warne, and Margot Hillel

The campaign against AIDS brought a new frankness to sex education in Australia, the argument being that only an informed population could protect itself from disease. This chapter is concerned with an earlier transformation, also inspired by the perceived threat of sexually transmitted disease, the debate surrounding the introduction of sex education during the fifty years prior to the outbreak of the Second World War. It documents a period in which the older moral discourse, which rested on concepts of abstinence and self-control, was displaced by a scientific one that aimed to show that public health education was as relevant for venereal diseases as it was for other feared maladies such as tuberculosis. But this new discourse was, we argue, equally conservative. Those organizations and individuals who favored extensive public education on the dangers of syphilis and other sexually transmitted diseases were determined to devise ways to spread public health messages without referring to complex aspects of both adult and juvenile sexuality that could have opened the way to a more liberal approach. The discursive shift that we trace here was not unique to Australia, but while Australian reformers drew freely on debates and materials developed overseas, they were applied to a self-consciously new nation, in which anxieties about race and morality were never far from the surface. Such anxieties increasingly served to locate the debates about sex education as central to national survival.

Among the earliest participants in these debates were Christian women's organizations. They took up the issue in the early 1890s after several notorious cases of rape and child sexual abuse heard in the courts alerted members to the fact that girls over the age of twelve were being successfully portrayed by their attackers as consenting adults. These cases, along with a perceived rise in the numbers of young women pregnant out of wedlock (Swain with Howe 4), led members of the Women's Christian Temperance Union (WCTU) to argue that ignorance was neither an assurance nor a safeguard of innocence and that girls and young women needed enough knowledge to protect themselves in the world around them (Warne 312). In the absence of locally generated sex-education resources, members of the WCTU, the Young Women's Christian Association (YWCA), and the White Cross League allied themselves with their parent organizations overseas and imported substantial numbers of pamphlets and other educational materials to distribute to the public. This strategy was not without risk. The 1896 Convention of the WCTU of New South Wales heard a number of people objecting to the nature of the pamphlets (WCTU of NSW 1896, 30), but the women persisted. They were still active during the First World War, slipping pamphlets into magazines sent to soldiers (WCTU of NSW 1915, 46).

Although they took the lead in addressing this issue, these reformers' presentation of certain elements of sex and reproduction was sanitized. Moreover, purity was to be reinforced by censorship of novels and motion pictures in order to restrict children's opportunities of acquiring sexual knowledge from other sources. This two-pronged campaign led both women's organizations and more official governmental health and educational representatives to juggle discourses that required both the suppression and the release of information. The evasiveness of their arguments and the aspects of sexuality that they refused to acknowledge created a situation in which the eroticizing of children in juvenile literature passed without comment, while factual information was often still buried in a mass of euphemism and obfuscation.

In the nineteenth century innocence was prized, leaving no space for the knowing child. Jill Bavin-Mizzi's analysis of Australian rape and carnal knowledge cases has confirmed the impression that so alarmed these Christian women. Juries found younger girls who described their experiences in "naive and descriptive terms" more credible than those who were able to explain sexual intercourse using adult terminology (Bavin-Mizzi 29). The girl who had the language to describe her abuse was condemned as too knowing to be an innocent victim. For instance, an eleven-year-old, accusing her father of sexual abuse in 1896, was able to give her evidence in a "remarkable intelligent fashion," but the doctor who "found her condition to be compatible with such an offence as the accused was charged with [added that] in reply to questions she gave him such answers as to lead him to believe that she was a precocious and knowing child and to throw considerable doubt on some of the occurrences which she described" (qtd. Scott and Swain 41).

In this climate, to give a child sexual knowledge was generally considered to be a danger rather than a protection, awakening instincts that could only lead to ruin. Addressing a 1916 conference organized by the Workers' Educational Association of New South Wales, Dr. H. Tasman Lovell warned his audience:

> Thoughts about, attention to, the instinct, its need, stimulus, or manifestations, automatically find a reverberation throughout the whole organism. . . . Muscles in various parts of the skeletal system, and in the visceral and reproductive organs, change their tone, and the whole series of changes is reported in consciousness as organic sensations, swelling the content of the mind as by an irresistible flood. Inhibition is then difficult in the extreme. Imagine what this must mean in respect of the sexual instinct, and then you may form some conception of the danger of turning the attention, not only to erotic thoughts, but to the sexual instinct at all. (46)

This distrust of even the most carefully framed information about sex rested on the fear that it would be almost impossible for the reader to separate such material from erotic or pornographic literature.

Accordingly, official literature produced for use with children, such as the 1911 *Teacher's Encyclopaedia* edited by A. P. Laurie, reflected the desire of the State Education Departments to steer classroom discussion well away from issues of sex. The fourth volume of this work was devoted to Child and School Hygiene and Child Physiology and Pathology but contained no mention of sexuality. The frontispiece, an illustration of a boy, included a flap that the reader could lift to see the workings of the internal organs, but the picture did not show the genitals. When, five years after the book's release, a deputation of moral campaigners and health educators approached the Victorian Minister for Education to request the inclusion of sex education in the curriculum, he adopted a similar approach, declaring that "it would be hard to get the children to listen in a reverent spirit, and there would be the danger of putting into their heads what was not there before, and that it would be difficult even for parents to speak to their children of tender years" ("Sex Hygiene" 13).

Like the authors of educational texts, writers of children's fiction often agreed that certain problems constituted scourges from which children needed to be protected. In Australian children's literature innocent unworldliness was usually assumed and sometimes celebrated. In *Five Little Bush Girls* (1918) by E. Lee Ryan, for example, the title characters live on an isolated farm with their parents. The farm is not doing well, and the children are well aware of their parents' economic difficulties. But they are innocent in sexual matters, knowing nothing of their mother's pregnancy until she returns home with the much longed-for baby boy. This ignorance is the more striking because it is at odds with the Romantic values emphasized in the book, which depicts the children as being one with nature and learning from it.

Constructions of such naïve, unworldly innocence persisted long after the First World War, highlighting the continuing contradictions inherent in society's views of what children should know. Sharing the concerns of censors and anxious to protect their young readers, publishers of children's books took pains to remove suggestions of supposedly questionable sexuality from their novels. Ward Lock Publishers regarded Ethel Turner's *The Little Larrikin* (1896), for example, as treading on "dangerous ground." Its potential immorality came from a subplot in which the older brother of the title character falls in love with a married woman. At the urging of her publisher, Turner changed this scenario to one in which the young man falls in love with the married woman's younger, unmarried, and therefore available cousin, preserving her readers from the danger of learning about a potentially adulterous sexual relationship. As Brenda Niall notes, because Turner accepted that "she was at her best in writing about children and therefore would be read mainly by children[, she] committed herself to observing certain limitations in subject matter and in language" (23).

Yet Turner did not completely turn away from depicting children in ways that we now consider to be sexualized and indeed potentially erotic; such depictions once again highlight contradictory notions of children and their sexuality. The critics who believed that children needed to be protected from the "corruption" of *The Little Larrikin* apparently failed to recognize the exploitation of the child image in other books such as Turner's *St. Tom and the Dragon* (1918). In this work Turner constructs nine-year-old Anne as innocent and therefore unaware of her own sexuality, but today's readers may well find the sexual content of the novel inescapable. Anne's abusive father, a widower, seems to be using his sadistic treatment of his daughter as a substitute for the legitimate sex he had with his wife. The language of the descriptions tends toward the graphic: "Last night he had caned her across the shoulders; the night before he had only caned her across the palms: there was nothing for it tonight but to tie her hands to a beam in the cellar below and whip her until her stupidity disappeared. . . . 'Down to the cellar and take your frock off,' he said thickly" (21).

When Anne flees this treatment, she is rescued by a whisky distiller and his wife, who first see her as a figure who combines dishabille with victimization: "a little girl in her petticoat, bare-legged, bare-armed, a bit of broken rope hanging from either wrist. A child with a frenzied face, and bleeding shoulders and bleeding arms" (45). The reader is being asked to sympathize with the child, seeing her as vulnerable because she lacks a protective mother, but is not being asked to identify the sexual nature of the violence. Her rescuers take delight in dressing her up, at least partly because of her body, since "beauty is a mighty stirrer of the heart, a magic solvent of iron, a real power in the world" (58). When her father is charged with beating her, Anne is required to give evidence in court. In the discussion between the police and the Warners as to how she should be dressed on this occasion, the power of physical appearance is again acknowledged. Also acknowledged, at least implicitly, are the concerns discussed at the beginning of this chapter regarding the ways that children's "knowingness" or naïveté may be expressed in their court appearances. Warner believes that Anne should be constructed as the beautiful innocent in need of protection—"the magistrate should have the opportunity to see that this was no child to be brought up by a drunkard"—while the policeman believes it would have more effect if she went to court in the "wretched clothes" she wore when she was first rescued in order to inscribe on her body the extent of her abuse (65).

The silences and dissemblings of children's literature were also evident in the approach of social purity campaigners. Speeches and pamphlets promoting caution among the young did so in such coded language that it was difficult for a youthful audience to discern just what they were being counseled to avoid. Directing their efforts toward concerned parents, social purity reformers warned that numerous aspects of modern culture were encroaching on children's innocence. Although most would-be reformers were reticent about how they had obtained sexual information in their own youth, they were free with their opinions about the way in which they believed others were told. American WCTU publicist Mary Wood-Allen, whose writings were widely used in late-nineteenth-century Australia, believed that the mothers of the current generation had "passed through [t]he[i]r development without instruction" (41), but few local campaigners were convinced. "In the poorer parts of all our thickly-populated suburbs," remarked one commentator in the feminist paper *The Australian Woman's Sphere* in 1900, "children . . . grow up with an intimate knowledge of all that is evil before they have a chance of knowing there is such a thing as goodness" (25), a view reminiscent of contemporaneous British assumptions about the link between overcrowding and sexual precocity in the working class. In the more genteel suburbs, warned speakers at the 1916 Workers' Education Association Conference, children were "educating themselves in a furtive and unwholesome manner" (Welsh 6), from "writings on walls in lavatories and fences . . . [and] overheard utterances" (Lind 77), or from "servants, or the 'bad' girls in a school" (Hinder 87). Knowledge gained in this way, often incorrect and always vulgar, uncouth, and coarse, cemented in young minds the impression "that secrecy and indecency are the necessary associates of sex" (Welsh 6).

This perceived link between "secrecy and indecency," together with the all-pervasiveness of sex, was what compelled these commentators to speak. While to modern perception their words sound like circumlocution, the reformers represented themselves as forthright. "Mothers generally are not acquainted with the fact that a terribly large percentage of Australian boys and girls of tender years in homes, schools and colleges, unconsciously suffer from sins that may be wisely, carefully prevented by them," Hannah Chewings warned fellow WCTU members in 1891 (80). "The evils cannot be remedied until they are courageously discussed," *The Australian Woman's Sphere* argued. "There is no greater power on the side of evil than the mock modesty which pretends that the wrongs from which some women suffer are too dreadful for other women to know about or discuss" ("Immorality" 25). And in 1892, Purity Departments within the WCTU in Victoria recorded that mothers' meetings were being "held weekly in the homes where mothers assemble, and are read to and addressed on this subject" (WCTU of Victoria 1892, 16).

Strong links between the White Cross League in Britain and the WCTU in the colonies of Australia propelled some women into active public education. Each WCTU state and local branch was meant to have a designated purity superintendent to coordinate and convince fellow members of the need for such activism. Working to distribute White Cross material, one WCTU member, recommending material written by an Australian (Lee) and an American (Bushnell), spoke glowingly of the possibility that Union members might overcome future problems by educating both young men and young women, albeit in a somewhat ad hoc fashion:

> I would like to recommend the "White Cross" series, they are splendid. There are a variety of leaflets, etc for Mother's meetings, also "Searets" [*sic*], "clean hands" etc by Kate C. Bushnell, M.D., "School-boy morality," "Shield and Buckler," etc and many others too numerous to mention here. There are some for men only; a good way to distribute these is to slip them in an envelope, and send them by post to any young (or old)

man you wish to influence by them. To encourage others, I will just mention that I have seen the good effects of some literature I have sent out, particularly, Mrs Harrison Lee's book on Marriage Laws. (WCTU of NSW 1893, 38)

By 1908 some educational circles began to advocate an enlarged scope for such work. Respected girls' school principal Elsie Morres, for instance, argued in a lecture delivered to a range of community groups during that year that sex education could be far more systematic than would be possible through the random distribution of pamphlets:

> A mother's duty in the early stage is in richly storing the child's mind with beautiful healthy precepts. . . . To have disciplined a child so that even at a very early age he looks on all that is dirty and disorderly and untidy, as shocking to his little sense of propriety, will be a long stride towards that health of mind which will be a coat of mail when the first suggestions of moral uncleanliness are presented to him. For dirt and disorder are the indication of a slothful and lazy disposition—a very fruitful soil for the germs of vice. . . . When a new life is introduced to the family, when the child is told of the advent of a baby brother or sister, that child will naturally set to wondering [about] the whole matter, and will surely ask an explanation, and if put off in a wrong, foolish, and to me, a wicked manner, by either father or mother, incalculable harm is done, instead of embracing a golden opportunity of sowing the first seeds of what might be a beautiful and moral character. . . . To sum up briefly then—
>
> Never deceive the child in its early days.
>
> Train the child in nursery, in school, in cleanliness and self-control—this latter being the keynote of all.
>
> Surround the child with an atmosphere of purity.
>
> Recognise the relationship of boys and girls. Treat them as rational thinking beings, and place before them the true ideals of refinement and purity. (4)

Christian women's organizations dedicated themselves to putting Morres's views into action. Arguing that innocence was being eroded, they insisted that it was the duty of the Christian mother to protect her child from misinformation. Angela Booth, a prominent independent advocate for sex education in Melbourne, argued: "The modern child cannot escape the conditions that keep its attention harmfully directed on matters of sex and its functions . . . this condition is not limited but runs through the whole fabric of modern life" ("Case" 162–63). In such a situation it was better for a child to receive accurate information about sex than to struggle on with half-knowledge. The role of the Christian mother was to come between the child and sexual danger, replacing the old shield of silence with a protective strategy of controlled information (Warne 322). Like British feminists of the 1890s, one YWCA speaker of 1912 urged mothers to reconceptualize their part in such education as a responsibility that would bring benefits of trust and closeness rather than an unpleasant task to be avoided:

> Nowadays, with so much helpful literature, a woman can hardly plead lack of knowledge of the way to give definite instruction to her children. The ways of Nature according to God's beautiful plan of life, can be simply and faithfully taught at the right time, and it is a daughter's right to expect such instruction from the mother instead of vague allusions that alarm rather than help. (Harris 13)

In order to prepare Christian mothers to perform this task, churches and women's organizations developed a range of resources that drew on metaphor and science to bring sex into discourse in a way that would not destabilize existing moral principles. Sex instruction focused initially on mothers, so that women who had received no accurate teaching themselves could face the task of teaching their own children with some confidence. The WCTU led the way, commencing classes for women in Victoria in 1892, importing pamphlets and books for distribution at branch meetings and public lectures.

Some prominent members of the women's movement were fortunate enough to be able to write and distribute their own opinions on the subject. They generated some of the first local sex-education material. At the forefront of these women, Maybanke Anderson, an outspoken advocate of women's access to contraception (the overturn of compulsory motherhood), White women's suffrage, and other civic rights, used her column in the Australian journal *Woman's Voice* to stimulate in her readers a sense of responsibility for the moral education of children (Roberts 171). Her columns, provocatively titled "Shall We Tell the Children?", "Shall We Confide in the Girls?", and "Shall We Talk to the Boys?" first appeared in 1895 and were soon republished as successful pamphlets.

Anderson was a committed educationalist who understood the value of instilling sound ideas in the minds of young children, but she had also learned from personal experience that where sex matters were concerned, more conservative members of the public soundly disagreed. A year before the publication of the sex-education pamphlets, she confided to Margaret Windeyer, a fellow leader in kindergarten circles, "As I daresay you know some of the specially particular people think I touch too plainly on sex matters." As a consequence, she commented, the special issue of *Woman's Voice* she was preparing in order to feature the Kindergarten Union would have sex matters "*rigidly excluded* except that there should be an open column letter bearing on the subject" (Anderson, emphasis in original). It is therefore not surprising that the educational techniques that she soon afterward promoted in her columns and pamphlets, particularly in the short story "Shall We Confide in the Girls?", were based on simple teachings about birds and flowers.

The YWCA and church-based organizations adopted similar techniques in their dual quest to educate parents and children alike; the Anglican bishops of Tasmania and Victoria specifically addressed their adult parishioners in *Help to Parents in Explaining Matters of Sex to the Young* (1918) while simultaneously acknowledging their debt to Mary Wood-Allen's earlier work *In her Teens*. The efforts of the Protestant churches lent the sex-education cause some gravitas but did not, in any real sense, overcome deeply rooted personal feelings of discomfort.

A far more pleasant tool was made available when Edith Howes, a respected New Zealand teacher and naturalist, published her lavishly illustrated and carefully constructed children's book *The Cradle Ship* in 1915. Here she created an enchanting world where an all-seeing mother closely supervised the workings of nature and morality. While the Australian adult population reeled at wartime public health revelations and wondered whether they should tell their children about venereal disease and "good-time girls," children born out of wedlock, and changing sexual behaviors, Howes combined in her story the ingredients thought necessary for an introductory prophylactic education. Billed as "a delightful fairy story for children [that] answers in a charming way, through flowers, insects, birds and animals, the question, 'Where does baby come from' " (Mothers' Union 1–2), it also gave sufficient imagery regarding the effects of venereal disease on individuals and communities to permit the story to resonate with parents as well as children. The Mothers' Union declared: "We can recommend it very heartily to mothers who find it is difficult to answer such questions" (Mothers' Union 2).

As a children's book and with its combination of information and whimsy, *The Cradle Ship* represented a new strand in the discourse. Compared to the modern, illuminating, and often sensational strategies that public lecturers were using to convince adult populations of the problems associated with venereal disease—one female speaker in New Zealand came to the notice of the Australian YWCA when she developed a strategy of projecting magic-lantern slides of babies deformed by congenital syphilis alongside slides of babies who were healthy, emphasizing as she did so that syphilis could damage the prospects of motherhood and deform newborn infants (Cockett 5)—Howes's style aimed to provide enlightenment in the gentlest possible manner. The healthy and beautiful imaginary world of "Babyland" had an abundance of "Fairy-mothers" among the trees and animals keen to exhibit the diverse methods of reproduction in nature.

Howes was careful to point out that the idyllic setting of Babyland was disrupted when some residents divorced sexual contact from reproduction:

> We take the very greatest pains to see that our father-stamens and mother-pistils stand where the bee who pushes in for honey must touch them, to cross the pollen from flower to flower, so that the babies may be strong. But many a bee is too lazy to enter by the front door and take the honey in the proper manner; she bites a hole in the honey spur at the back of the flower and steals the honey. She never comes near the pollen. The honey is there to be given her, but she would rather steal it than go to a little trouble to get it. (Howes 25)

The resulting community in the book, diseased and deficient, with flowers so "twisted and badly shaped that they will never open, and their seed-eggs must die," was known by the other residents of Babyland as "That Dreadful Garden." The "Good" fairy mothers complain that their own proximity to the garden has endangered their families, since

> Insects coming from that garden bring bad pollen to our flowers. Our babies will be ill and die, or they will grow up poor stunted things we shall be ashamed of. Indeed if things go on as they are, you spurned flowers may well be glad if no bee enters by the front door to bring pollen from such an evil place. Something must be done. We must protect our babies. (Howes 27)

To temper the shock that the image of the blighted garden may have caused her childish readers, Howes continues the story to show the possibilities of maternal activism: " 'I should be ashamed to own such neglected children,' said Mother quietly. She stood up very straight and tall, and noticed everything in the garden: babies shrivelled and starved, babies sick and dying, babies dead and mildewed. The wonderful look came into her eyes, and the fairies shivered in fear, for they felt her mother power" (Howes 28).

In the eyes of the international women's movement, White women in Australia and Maori and Pakeha women in New Zealand had, in fact, achieved an impressive civic elevation in having gained the vote by 1902. One Western Australian parliamentarian, Henry Kenny, had proudly declared in a moment of unguarded parliamentary debate over the issue in 1898,

> It is admitted on all sides that, while men are only the progenitors of our race, the women are its saviours; and that on the future of the Anglo-Saxon race to which we are all proud to belong, and on the future of the civilised races of the world, women are exercising a higher influence and playing a more important part and will continue to do so, than men can aspire to do. On these lines I claim the right of woman to have the vote. (Qtd. Grimshaw and Ellinghaus 12)

Such racialized, gendered language permeates the redemptive message of the story in which Howes promises that "good" mothers will improve the "bad."

The stress on the power of good mothers provided women with strong encouragement to adopt the reforming zeal necessary to "save" Australian society. But "mother power" was based on the premise of women disciplining other women, devaluing some women while elevating others (Ross 199–216, Yeo 145). In Australia, while White women were being encouraged to see themselves striving to be the "saviors" of the "Anglo-Saxon race," Indigenous women were facing child removal laws that separated generations of Indigenous children from their mothers (Commonwealth of Australia). In fact, women authors such as Howes were proud to show that White women had advanced to the point of being able to handle the most sensitive and subtly encoded aspects of modern society, even explaining matters of sex to the young, while during this crucial period state and federal governments chose to see Indigenous Australian mothers as unable to look after even the basic needs of their children.

The Cradle Ship therefore acknowledges the struggle over the ownership of "proper" sexual knowledge raging in the White community (only alluding to conditions in other communities), even as it portrays the debate about sex education in a simplified way that children in Australia and New Zealand could understand. Both the grandmother, who tells the children not to be rude when they ask where the new baby came from, and "Mr. and Mrs. Darken," who try to stop the family from going on their mission to Babyland, are described as "ignorant people, who think it wrong that children go where they did not go when they were children" (Howes 10–11). The modern mother, by contrast, is encouraged to eschew the sensational moral hyperbole of the Victorian era by adopting a language of "birds and bees" to explain to children the complex issues of sex. Procreation, at least, is to be seen as natural, and mothers are encouraged to see their own maternity as an avenue for explaining the basic facts about sex and reproduction to their children.

This carefully coded approach to sex education drew an enthusiastic response. "There are few little boys or girls who will not be infinitely touched and refined by the knowledge that their mother bears within her, and nourishes, the baby that is to come into the home," wrote one YWCA member (Ellis 8–9). Only the communication of such "balanced" truth could provide a significant counterweight to the grandstanding shock tactics that dominated public discourses about sex as competing discourses increasingly expressed anxiety about rates of premarital pregnancy and venereal disease. Members of the women's movement eagerly countered the smut associated with talking about sex by positioning themselves as progenitors of Anglo-Saxon "mother power."

The Cradle Ship and later Australian books such as Irwin Smith's *The Story of Ovum and Sperm; and How They Grew into the Baby Kangaroo*, published in 1920, and Marion Piddington's *Tell Them! Or the Second Stage of Mothering: A Handbook of Suggestions for the Sex Training of the Child*, published in 1925, provided mothers with the materials they needed to undertake this educational role, entertaining and amusing didactic tools that did not dwell unduly on medical issues or on sin. Kerreen Reiger has argued that this new openness toward sex education needs to be seen in the context of wider concerns about regulating the health of the population, concerns shared by health reformers, committed to enlightenment through scientific knowledge, and conservative religious groups, who wanted to reform both private and public morality. But while both groups condemned the previous silence on sexual matters, they struggled to find a common response (Reiger 179). In addition, there was public sentiment to overcome. Campaigns to introduce sex education into schools prior to the First World War had been unsuccessful. The Rev. William Bligh of the White Cross League was banned from state schools in Victoria after parental complaints that his lectures, designed to inform children of the dangers of premature sexual activity, were too explicit (Reiger 182–83). Nor was the Education Department persuaded by calls from the state's chief justice, Sir John Madden, for an alliance between "the schoolmaster . . . the pastor and the physician" to counteract the "false prudery [that] has kept all references to physiological dangers out of school instruction" (untitled editorial 2).

But the concern about venereal disease that reached a peak during the First World War gave renewed impetus to the debate. Although the existing alliance between reformers and conservatives was maintained, it was the conservatives who were able to take control of the discourse, setting a pattern that would shape sex education for the next fifty years. Modernity, they argued, fundamentally undermined the innocence of childhood, creating an environment in which only the informed could make the choice to remain pure. Condemning the discursive restraints of past eras as counterproductive in the modern world, they nevertheless argued that high moral standards had to be maintained if society was not to crumble. Piddington, writing in 1925, put this new argument well:

> The old order of ignorance said, "You must not." "It is wicked." "God will punish you!" The new order of enlightened thought teaches, "You will understand why you should not"; "you will hurt others"; "you will punish yourself." Thus the new teaching, in contrast with the old preaching, will secure conformity with the laws of God through intelligent understanding of them, the only source from which permanent benefit to mankind may be derived. (21)

While Jeffrey P. Moran has argued that in Chicago, Progressivist organizations "deeply threatened by evidence that individuals and institutions were failing to rebuff the modern city's temptations . . . were confident that they, as an enlightened elite, had the ability to stop the decay" (482), in Australia this self-identified "enlightened elite" took a more conservative line. Although it is easy to emphasize the dramatic nature of the discursive shift, contrasting the morality of the old purity movement with the progressive attitudes of the new, to do so is to miss the continuity in underlying goals. Jim Jose, in examining the South Australian debates, characterizes this shift as a "regrouping of moral sentiments within a discourse informed by medical science, a convergence whereby the earlier configurations of ideas about sexual and racial purity were recuperated within a medical regime of knowledge . . . [in which] ideas of sexual and racial purity remained embedded" (47). The racial imperative that underlay the eugenics movement in Australia made what was a major shift less controversial because it could be argued that it served the greater good, ensuring that the young would remain healthy enough to produce the next White generation.

There is a sense in which the 1916 Workers' Educational Association Sexual Hygiene Conference (cited earlier) captures this moment, the point at which, in Foucauldian terms, sex was brought into discourse in order to more effectively contain or control it (Foucault 11). The purpose of the conference, delegates were told, was to find effective means "to train the children to grow up pure in mind and in body, fit citizens with whom to entrust the future of the race." The key reference point was a statement of what Michel Foucault would later describe as the repressive hypothesis:

> Our mental attitude towards the problems of sex and venereal disease has undergone an evolution. The open licence of the Restoration and of the

> Regency led to a reaction which went too far. It developed the prudery of Victorian times, which was offensive to the truth and harmful to the race. We are now entering upon a more promising attitude of candour and common sense. (Welsh 2)

"The policy of silence has failed disastrously and must be abandoned," Bishop Long declared, but the policy of abstinence was to remain intact. "No compromise is permissible with regard to the universal obligation of the law of purity. Sexual intercourse other than as sanctioned in wedlock we must unflinchingly denounce as sin," he continued (106). And Dr. Tasman Lovell, an educator, added, "It would be a sad and fatal thing if efforts in this direction were to awaken thoughts of sex without ultimately tending to provide the individual not merely with reasons for inhibiting, but with the actual power to inhibit" (50).

In answering the question "Why teach?" delegates to the conference reiterated the arguments used by earlier reformers. "To send our boys and girls, ignorant of sex, out into the world, is like sending an army out to fight badly armed and inadequately equipped," said the New South Wales Minister for Education, the Honorable Arthur Griffiths (38). Booth warned, "The boy and girl, equally ignorant, hear nothing but the call of the sex-instinct" ("Prostitution" 31). "So long as we allow our children to learn of this subject from vicious sources, so long will uncleanness cling to it," Teachers' College principal Dr. J. Smyth added ("Standard" 195).

There was less unanimity as to what should be taught. While there was general agreement that children needed to be prepared for the onset of puberty, there was less certainty as to what exactly this preparation would involve. "What we should teach is the hygiene of the sexual organs, and not the physiology," Sir Thomas Anderson Stuart, dean of the Faculty of Medicine at the University of Sydney, declared (91). Smyth disagreed:

> Merely to talk about the good, the pure, and the sacred, without giving content to these ideas, is useless. A boy must be told about the most important organs in his body, must have conveyed to him definite information about birth and reproduction, must learn the connection between the semen and virility of the body, mind and soul, and yet must have all this conveyed to him so that he will feel the responsibility of the trust of possible fatherhood committed to his care, will reverence women, and will understand that it is in his thoughts and acts in regard to this subject that he can best prove his manhood. ("Education" 167)

But the most vigorous debate at the Conference centered on the question of who should teach, with each of the different professional groups represented attempting to control the discourse. Ideally, all agreed that sex education would be provided by the parent within the context of a loving family environment, but few of the professionals had faith in the ability of parents to undertake this task. "It is a task for the tact and understanding of the parent," Bishop Long declared. "Mothers and fathers should know . . . what they are to teach, and how to teach it . . . [but] it will be necessary to put the very words in their mouths if they are not to do more harm than good" (101).

For feminist Jessie Street this was a temporary, or generational problem:

> The children of today of course are the parents of tomorrow, and as it is so difficult for us to teach parents who are fixed in their ideas and who have been brought up with the ideas of mock modesty, responsible for present day conditions to a great extent, the question is how we can teach them, and if we cannot teach them how are they to teach their children. Is it not better for us to endeavour to work out some scheme by which we can teach the children so that they will be so impressed with the necessity of knowledge that they in turn will pass on the teaching to their children? (74–75)

For Smyth, however, the damage was done, since "The school, by compulsory attendance, by home lessons, and by other demands, has interfered with home training and education. The school should, therefore, make amends by giving due attention to moral training, and this includes definite instruction in sex" ("Education" 172). And one classroom teacher, Miss Hinder, gave a practical example of how the subject could be raised: "When you get girls sufficiently interested to ask about what is the male and the female of plants, they will go away and do thinking about it, which will help them more than anything merely told to them in a lecture" (88).

The debates and opinions aired at the conference were being echoed in society as a whole. Pressure increased on state education departments during and after the war, and some responded by introducing limited instruction in teachers' colleges. In Victoria in July 1916, a delegation from the Society for the Promotion of Social Hygiene called on the Victorian Minister of Education, Sir Harry Lawson, to recognize "that the ravages of venereal disease were so great that the teaching of social hygiene was imperative . . . [and to authorize] the distribution of literature dealing with sex hygiene" in schools ("Sex Lessons" 6). A prominent women's columnist, E. A. Allen, who wrote under the pen-name "Vesta," commented: "The time has come when this work must be extended in order to reach the mothers of the girls in our secondary schools, and the members of Mothers' Clubs in our secondary schools." To further confirm the probity of the operation, Allen assured Australian readers that similar strategies were being implemented in England "in the hope that this will be the last generation of women to shirk the duty of instructing their children" ("Sex Hygiene" 13).

Initially the Minister proved reluctant to act, citing the "danger of putting into their heads what was not there before" and the difficulty "for parents [in speaking] on these subjects to children of tender years" ("Sex Hygiene" 13). Yet Reiger has found evidence that in the interwar years individual schools did introduce sex education, integrating the material into the standard curriculum rather than making it a separate subject with outside instructors (183). Beginning in the 1920s the small Victorian Sex Education Society offered training courses to mothers, teachers, and club leaders on biological, moral, and spiritual issues (184). Although the content of such courses professed a rational rather than a moral approach, Reiger has argued, the moral exhortations remained, now with a psychological overlay. Information was provided in order to persuade or frighten the child into accepting abstinence (186).

Other educational and medical experts showed a continuing distaste for the idea of stepping into the mother's shoes. In 1929, for instance, a national conference of schoolteachers held in Perth rejected the idea that anyone other than parents could teach sex education. Dr. Roberta Jull, author of *Sex Physiology*, spoke at the conference, arguing that "parents must purify and simplify their own attitude of mind and so be fitted to guide and direct their children" (Rischbieth Papers MS 2004/12/1650). To aid others in this endeavor, she delivered a series of four lectures on facts about sex to girls and their mothers in conjunction with the Women's Service Guild of Western Australia, attempting thereby to coordinate the talents of the mother and those of the health expert (Rischbieth Papers MS 2004/12/1658).

Other schemes that sought to mobilize an informed motherhood were less successful. In the early 1930s the Women's Service Guild tried to obtain permission to have sex education given to mothers in Western Australian Infant Health Centres so that they in turn could educate their children from an early age. Members of the Women's Service Guild argued that since 85 percent of mothers attended these centers, they were an ideal avenue for the education of women in modern scientific ideas about sex and reproduction (Rischbieth Papers MS 2004/12/1704). Though the idea of teaching sex education continuously from infancy was well established by the late 1920s, Dr. Strang, the Medical Supervisor for the Infant Health Centres, refused to cooperate with the plan, arguing that the Women's Service Guild was aiming at the wrong audience: "I think that the mothers who attend the Centres would not be the ones who would take a vital interest in this subject. I think rather that it would be wiser to get in touch with mothers whose children are older, that is, about 12 years of age" (Rischbieth Papers MS 2004/12/1704). This growing focus on adolescence in the 1930s marked a rapid but ambivalent shift from mothers teaching about sex gradually in the home, to teenagers learning about sex and human development in a public, although not strictly educational, sphere. It was, perhaps, also an admission that the strategy of early education had failed.

The reformers who sought to control the flow of information about sex and to maintain childhood as a period of "informed" innocence were well aware of the countercurrents in both literature and popular culture that sought to sexualize the child. Rather than contribute to such tendencies by using explicit language or relaxing, in even the slightest degree, their insistence on abstinence, they adopted an oppositional stance, urging parents and teachers to mobilize against what they perceived as the evils of modernity. They advocated sex education not to enlighten young people but to protect them, drawing upon metaphors of war and massacre to stress the urgency of their cause. Their motivation for so dramatically bringing sex into discourse was to control, or at the very least contain, a force that they saw as threatening both personal and national survival.

While we have argued that, in the period under discussion, such conservative forces were successful in taking control of the debate, defining the limits of acceptable discourse, and developing a new vocabulary in which sex could be more openly discussed, there is little to suggest that children were substantially enlightened as a result of these efforts. The new books and pamphlets may have eased the embarrassment of both parents and teachers, but they did little to help children decode the meaning of the schoolyard rhymes that remained the most common means of transmitting sexual information. Yet in a sense the two discourses were mutually constitutive, the crudity of the schoolyard parodying the evasiveness of the flower and fairy metaphors of the sex-education literature, similarly burying its meaning beneath a complex language accessible only to the initiated:

It's only human nature after all
To get your favourite girl against the wall,
And let your accumulation
Meet her accommodation,
And increase the population
Of the coming generation. (Turner, Factor, and Lowenstein 118)

Works Cited

Anderson, Maybanke. Letter to Margaret Windeyer, 7 November 1894. Margaret Windeyer Papers, Mitchell Library, Sydney.

Bavin-Mizzi, Jill. "Understandings of Justice: Australian Rape and Carnal Knowledge Cases, 1876–1924." *Sex, Power and Justice: Historical Perspectives of Law in Australia*. Ed. Diane Kirkby. Melbourne: Oxford University Press, 1995.

Booth, Angela. "The Case for the Sex Education of the Adolescent Girl." Workers' Educational Association of New South Wales. 159–65.

——. "Prostitution." Workers' Educational Association of New South Wales. 14–39.

Chewings, Hannah. "Purity Superintendent's First Report." *Report of the 3rd Annual Convention of the Woman's Christian Temperance Union of South Australia*. Adelaide: Woman's Christian Temperance Union of South Australia, 1891. 80–81.

Church of England in Australia. *Help to Parents in Explaining Matters of Sex to the Young, Authorised by the Bishops of the Church of England in Australia and Tasmania for the Use of Parents, Guardians and Teachers*. Sydney: William Andrews Printing, 1918.

Cockett, Rev. C. Bernard. "Educational and Moral Values." *The Association Woman* (November 1917): n. pag.

Commonwealth of Australia. *Bringing Them Home: National Inquiry into the Separation of Aboriginal and Torres Strait Islander Children from Their Families*. Sydney: Stirling Press, 1997.

Ellis, Dr. Constance. "Woman after the War." *The Australasian Women's Quarterly* (July 1915): 8–9.

Foucault, Michel. *The History of Sexuality, Volume 1: An Introduction*. Trans. Robert Hurley. London: Allen Lane, 1978.

Griffiths, Arthur. "Opening Address." Workers' Educational Association of New South Wales. 38–39.

Grimshaw, Patricia, and Katherine Ellinghaus. "White Women, Aboriginal Women and the Vote in Western Australia." *Women and Citizenship: Suffrage Centenary*. Studies in Western Australian History no. 19. Ed. Patricia Crawford and Judy Skene. Perth: University of Western Australia Press, 1999. 1–19.

Harris, E. B. "Mothers and Daughters—A Plea for Closer Comradeship." *The Australasian Young Women's Christian Association Quarterly* (October 1912): 12–13.

Hinder, Miss. Contribution to "Questions and Discussion Following the Reading of Dr. Lind's Paper on 'Sex Irregularities of Childhood and Youth.' " Workers' Educational Association of New South Wales. 86–90.

Howes, Edith. *The Cradle Ship*. Melbourne: Cassell, 1916.

"Immorality or Prudishness?" *The Australian Woman's Sphere* (November 1900): 25.

Jose, Jim. "Sex Education, the Family and the State in Early Twentieth Century South Australia." *History of Education Review* 27.1 (1998): 33–52.

Laurie, A. P., ed. *The Teacher's Encyclopaedia of the Theory, Method, Practice, History and Development of Education at Home and Abroad*. Vol. 4. London: Claxton, 1911.

Lind, W. A. T. "Sex Irregularities of Childhood and Youth." Workers' Educational Association of New South Wales. 77–91.

Long, George Merrick. "How the Church Could Co-operate in the Teaching of Sex Hygiene." Workers' Educational Association of New South Wales. 105–22.

Lovell, H. Tasman. "The Child and the Sex Problem from the Psychological Point of View." Workers' Educational Association of New South Wales. 40–76.

Moran, Jeffrey P. " 'Modernism Gone Mad': Sex Education Comes to Chicago, 1913." *Journal of American History* 83 (September 1996): 481–513.

Morres, Elsie. "Moral Training of the Young." *Geelong Advertiser* (June 20 1908): 4.

Mothers' Union of Australasia. "Editorial." *Mothers in Australia* 3.4 (1920): 1–2.

Niall, Brenda. *Seven Little Billabongs*. Melbourne: Melbourne University Press, 1979.

Piddington, Marion. *Tell Them! Or the Second Stage of Mothercraft: A Handbook of Suggestions for the Sex Training of the Child*. Sydney: Angus and Robertson, 1925.

Reiger, Kerreen. *The Disenchantment of the Home: Modernizing the Australian Family 1880–1940*. Melbourne: Oxford University Press, 1985.

Rischbieth Papers (Bessie), National Library of Australia:

MS 2004/12/1650. Unsourced news clipping dated 1 April 1929

MS 2004/12/1658. Draft for Advertisement in the *West Australian*, 31 May 1929.

MS 2004/12/1704. Letter from Miss Cass to Dr. Strang, 11 April 1932.

MS 2004/12/1704. Letter to Miss Cass from the Medical Supervisor of Infant Health Centres, WA, Dr. Strang, 26 April 1932.

Roberts, Jan. *Maybanke Anderson: Sex Suffrage and Social Reform*. 1993. Marrickville, N.S.W.: Ruskin Rowe, 1997.

Ross, Ellen. "Good and Bad Mothers: Lady Philanthropists and London Housewives before World War I." *Gendered Domains: Rethinking Public and Private in Women's History*. Ed. Dorothy O. Helly and Susan M. Reverby. Ithaca: Cornell University Press, 1992. 199–216.

Ryan, E. Lee. *Five Little Bush Girls*. Melbourne: Australasian Authors Agency, 1918.

Scott, Dorothy, and Shurlee Swain. *Confronting Cruelty: Historical Perspectives on Child Abuse*. Melbourne: Melbourne University Press, 2002.

"Sex Hygiene: State Minister's View." *Argus* (12 July 1916): 13.

"Sex Lessons in Schools Refused." *Geelong Advertiser* (12 July 1916): 5.

Smith, Irwin. *The Story of Ovum and Sperm; And How They Grew into the Baby Kangaroo, Stories of Birth and Sex for Children. Part II: The Story of Sex*. Sydney: Australasian League of Honour, 1920.

Smyth, J. "Sex Education for Boys." Workers' Educational Association of New South Wales. 166–84.

Smyth, J. "The Double Standard of Morality." Workers' Educational Association of New South Wales. 185–98.

Spurr, F. C. *The Red Plague: A Lecture Delivered by the Reverend F. C. Spurr at the Town Hall, Melbourne on 25 September 1911.* Melbourne: J. Kemp, 1911.

Stuart, Thomas Anderson. "How Much Should Be Taught to Children and Young Adolescents." Workers' Educational Association of New South Wales. 91–104.

Street, Jessie. "Questions and Discussion Following the Paper by Dr. Lovell, on 'The Child and the Sex Problem from the Psychological Point of View.' " Workers' Educational Association of New South Wales. 72–76.

Swain, Shurlee, with Renate Howe. *Single Mothers and Their Children: Disposal, Punishment and Survival in Australia.* Cambridge: Cambridge University Press, 1995.

Turner, Ethel. *The Little Larrikin.* London: Ward Lock, 1896.

——. *St. Tom and the Dragon.* London: Ward Lock, 1918.

Turner, Ian, June Factor, and Wendy Lowenstein, eds. *Cinderella Dressed in Yella: A Collection of Australian Children's Play Rhymes.* Melbourne: Heinemann, 1969.

Untitled Editorial. *Geelong Advertiser* (30 August 1909): 2.

Warne, Ellen. "Sex Education Debates and the Modest Mother in Australia, 1890s to the 1930s." *Women's History Review* 8.2 (1999): 311–28.

Welsh, D. A. "The Massacre of the Innocents." Workers' Educational Association of New South Wales. 1–14.

Women's Christian Temperance Union of New South Wales. *Report of the 11th Annual Convention of the Women's Christian Temperance Union of New South Wales.* Sydney: Women's Christian Temperance Union of New South Wales, 1893.

——. *Report of the 14th Annual Convention of the Women's Christian Temperance Union of New South Wales.* Sydney: Women's Christian Temperance Union of New South Wales, 1896.

——. *Report of the 33rd Annual Convention of the Women's Christian Temperance Union of NSW.* Sydney: Women's Christian Temperance Union of New South Wales, 1915.

Women's Christian Temperance Union of Victoria. "Heredity-Hygiene Department Report." *Annual Report of the Women's Christian Temperance Union of Victoria.* Melbourne: Women's Christian Temperance Union of Victoria, 1892. 16.

Wood-Allen, Mary. *Teaching Truth.* Melbourne: Signs Publishing, 1892.

Workers' Educational Association of New South Wales. *Teaching of Sex Hygiene: Report of a Conference Organised by the Workers' Educational Association of New South Wales and Held in the Union Hall, Sydney University on November 23, 24 and 25, 1916.* Sydney: Burrows, 1918.

Yeo, Eileen Janes. *The Contest for Social Science: Relations and Representations of Gender and Class.* London: Rivers Oram, 1996.

Chapter 3

Eyes Tightly Shut, Lying Rigidly Still, and Thinking of England? British Women and Sex from Marie Stopes to Hite 2000

Lesley A. Hall

One of the most interesting questions about literature written with a didactic intention is the extent to which that intention was actually accomplished. In a trenchant article on advice to mothers, Jay Mechling has argued that didactic literature may have little to do with behavioral patterns and their modification. While the complete lack of any connection that he posits has been contested (Porter and Hall 218–19, 283), certainly the relationship between works of sexual pedagogy and changes in sexual attitudes and behavior is far from straightforward. This chapter looks at one particular strand within the genre of sex-advice literature in Britain in the early twentieth century, and compares its messages with the experiences reported in a number of social surveys concerned with sexual attitudes and behavior—themselves, of course, somewhat problematic.

Early in the year 2000, a new edition of Shere Hite's famous report on women and sex was published and its findings about British women extensively discussed. As Anthony Browne commented in the Sunday broadsheet *The Observer*, Hite claimed that British women "have got more out of the sexual revolution—and have less reservations about it" than women in the United States, Australia, and New Zealand, being "self-confident about sex" and "at ease with their bodies." Although "they are far less satisfied with their partners than women elsewhere," in stereotypical Brit fashion "British women show a remarkably stoical—even tolerant—attitude towards their men." Could this be taken at face value, it would seem that British women have come a long way since 1900.

It is hardly necessary to recount the well-known if probably chimerical clichés of Victorian sexuality: the dictum that "ladies don't move"; the bride found by her new-wedded husband chloroformed and with a note pinned to the pillow, "Mamma says you must do as you like"; and of course the advice to "close your eyes and think of England." Recent historiography has done much to deconstruct this simplistic blanket characterization of Victorian sexual life, presenting a far more complex reality. But a totally revisionist view of Victorian sexuality as happy and fulfilling is hardly possible to sustain. There was silence, suffering, anxiety, and fear enough.

This chapter considers works of sexual (specifically, marital) advice written by a group of women, feminist in their politics, post-Victorian in their self-positioning, who saw it as their destiny to overthrow the barriers to women's right to sexual pleasure. It then turns to various observations and surveys from the 1920s to the 1950s that tried to ascertain the ordinary woman's feelings about sex. Some recent material on the sexual attitudes and experiences of British women provides helpful strategies for exploring the disjunction between the very different attitudes toward sexuality manifested by these two groups of women, an articulate minority and an often inarticulate majority.

My title identifies Marie Stopes as a beginning point, but it is necessary to pay some brief attention to earlier debates among British women on sexual issues, which formed a significant part of the context from which she emerged. From the late nineteenth century, and stimulated by the fervid campaign against the Contagious Diseases Acts (among other causes), women had been critiquing the sexual system in existence. This system was predicated upon a double moral standard, which regarded a lapse in chastity as a fatal fall for a woman but a peccadillo in a man. The double standard was manifested in many ways: in legal proceedings, medical discourse, and general social attitudes. It can be seen in the terms of the 1857 Divorce Law, which made a wife's adultery grounds for divorce while providing that a woman could not divorce an adulterous spouse unless he had committed an additional matrimonial offense, namely cruelty, desertion, sodomy, or incest. While recent historians have indicated some of the ways in which the double standard was moderated in practice—the sympathy that might be accorded to a woman who had "fallen" once to the seductions of a vile cad, the juridical interpretation of "cruelty" to include knowing transmission of venereal disease—nonetheless this was the pervasive way of thinking about the relations between the sexes. Late nineteenth- and early twentieth-century women who broke the rule of silence about sexual matters that social custom imposed on women spoke out about the very considerable downside of sex for women of the time: cruelty, callousness within marriage, and the venereal diseases constantly being circulated by male promiscuity. Lust was male, and lust was evil.

Nevertheless, a few women were beginning to query the idea that sex, per se, was necessarily a bad thing, to which any woman would only submit at wide intervals in order to have children. There were the much-cited debates in the maverick feminist journal *The Freewoman* on whether women were, unlike men, innately chaste. Similarly, the provocative articles of Rebecca West and the writings of the radical socialist feminist Stella Browne suggested that if the existing state of society, which promoted men's sexual exploitation of women and the penalizing of women who transgressed convention, were to be overthrown, sex might be very different (Bland; Hall, "Generation").

The Sexual Rights of Women

While West and Browne are recognized as subversive and radical voices, the radicalism of the women who were writing sex manuals in the early twentieth century has often been overlooked because their new vision of female (and concomitant male) sexual pleasure was firmly grounded in the context of marriage. Their prescriptions were said to be for the improvement of married life and to stabilize unions, at a time, just after the First World War, when society was believed to be in moral upheaval and the institution of wedlock in profound jeopardy. Still, it would hardly have been possible to write as they did for a broad popular audience except by making this concession to conventional mores. They had an acute sense of the limits within which they could safely and effectively work, as indeed did other commentators on sexuality during this period; it may be noted that contemporaneous women writers who were advancing arguments for a more liberal view of erotic relationships beyond the standard parameters of marriage did not combine these arguments with practical suggestions for achieving erotic bliss. All had to make some kind of compromise to get their messages into the public domain and elude the heavy hand of censorship.

Stopes (1880–1958), the best known of the women discussed here, is probably chiefly remembered for her crusade for birth control; she believed strongly that women of all classes should be able to space their pregnancies for the good of their own health and the better upbringing of the children they did have. But she also believed that women had both the capacity and the right to achieve sexual pleasure, and should experience motherhood as a wanted joy rather than an imposed burden. She took the ideas ventilated in marginal and radical venues and circles and turned them into a domesticated vision of the possibility of sexual bliss within society as it already existed (though she strongly implied that the transformed marriages thus created would themselves be an agent of social change). She was undoubtedly a feminist. Her first publication in the marital advice genre and the most famous of her writings (she had previously written scientific papers and works of popular science, as well as plays, short stories, and poetry), *Married Love* (1918) was a passionate plea for the recognition of women's right to sexual pleasure—within marriage, naturally. Stopes emphasized the need for a woman to be aroused as a preliminary to intercourse, and for satisfactory orgasm for both parties (Hall, "Science"). She also propounded the somewhat more controversial theory that women were subject to "sex-tides" related to the menstrual cycle, and that the apparent "contrariety" of women was due to this fluctuation of their desires (Stopes, *Married*, 18–49). This contention was one way of conceptualizing female sexuality as not merely a reciprocation of male desire.

Stopes became practically the brand name for this new genre of sexual advice, but she was not the only woman writing in this field. The woman doctor Isabel Hutton seems independently to have been inspired to produce *The Hygiene of Marriage* (1923) by the inquiries she received in her medical practice, and possibly by the experience of adjustment in her own relatively late marriage to a career Army officer. Like Stopes, but in more sober tones, she indicated the importance of female orgasm ("essential to the health and happiness of the couple" [*Hygiene* 46]), and the need for preliminary arousal before intercourse was attempted. She mentioned that "the clitoris . . . has, indeed, more to do with the arousal of sexual feeling than any other area of the female body" (51), and recommended, should a woman not experience a climax during intercourse, that her husband should "attempt to complete the act . . . by digital stimulation" (67).

Another woman doctor, Helena Wright, produced a short guide, *The Sex Factor in Marriage*, in 1930, and a sequel, *More About the Sex Factor in Marriage*, in 1947. She placed supreme importance on the role of the clitoris in female arousal and satisfaction. Although we often assume that Freudian concepts of the supremacy of the "vaginal orgasm" and the immaturity and general undesirability of the clitoral version were paramount at the time when she was writing (something probably much truer of North America than the United Kingdom), Wright was adamant that a penis-in-vagina model of intercourse, though it satisfied male needs, was not adequate for many women. She strongly advocated women familiarizing themselves with their genitalia and finding out the best ways of stimulating themselves (*More*, 58–68).

Finally, another physician, Joan Malleson, whose *Any Wife or Any Husband* (published under the pseudonym "Medica" in 1950) was based on her pioneering sex therapy work, was perhaps less radical than Wright. On the one hand, she appeared to believe in the existence of the vaginal orgasm, although she conceded that the most common complaint she came across among women was "vaginal anaesthesia." Nonetheless, she strongly argued that women who were unable to achieve this mode of orgasm would obtain a high degree of satisfaction from clitoral orgasm—which was infinitely preferable to striving after an unattainable ideal (58–64). Slightly younger than Wright, she may have been more influenced by the popularized Freudian notions about vaginal orgasm, but does not seem to have considered the clitoral variety inferior. Rather, it was a perfectly satisfactory outcome.

In short, Stopes was by no means unique: there was a tradition of women writing popular guides to sexual pleasure within marriage from a female-centered point of view, emphasizing the particular nature of female pleasure and indicating the importance of the clitoris. These books were well known within the family planning and marriage guidance movements and influenced the ideas of those working in these fields.

Moreover, the idea of the desirability of female sexual pleasure within marriage was becoming increasingly disseminated, for example through women's magazines, which reached an even larger audience than the sex manuals. By the late 1930s the problem pages and advice columns, at least, were conceding, even advocating, this stance. A January 1938 article by Christine True, "our Human Problems Expert," in *Mother: The Home Magazine* declared that "the sex act [is] a normal and natural expression of love" and advised that "the feeling of The Song of Songs should be present in marriage" (19). Later the same year True responded to one of her female correspondents by suggesting that her unnamed problem was due to living in "abnormal circumstances. . . . You are a normal healthy woman and it is natural for you to feel as you do" and making what appears to be a covert offer of birth-control advice—"information that would make it possible for you to lead a perfectly normal married life, without the risk that the doctor said you must not take" (April 1938, 118). To a woman who complained that she was "terribly disappointed" that sex had "all been so unsatisfying. . . . my husband knows very little about this side of marriage, and I cannot help resenting bitterly that he dared to marry without knowing how to make a woman happy in this way," True promised to recommend books, advising her also to talk to her husband (May 1938, 110). Similarly, in 1941, "The Family Doctor" in *Lucky Star* counseled the wife in a couple who had not been living "as husband and wife" for fear of having a baby in wartime that "the kind of life you are now living is not good for either of you" ("Ask," January 1941, 13). In January 1945 Ray Allister, the "agony aunt" of *Modern Woman and Modern Home*, cautioned a reader who was distressed at her physical indifference to her fiancé: "It seems to me as dangerous to marry without *any* plain sex desire as to marry merely because of it" (83). And in the same year, the "Heart to Heart" column in *Woman's World* advised an inquirer: "If you are courting seriously there is no reason why you should not respond to his kisses" rather than remain indifferent (11).

To be sure, much work remains to be done on the analysis of women's magazines: not only did different sections of any single issue of any particular magazine convey different messages (problem pages versus fiction, e.g.), but different magazines also had different moral stances and were addressed to different intended audiences, both in age and class, and in some cases also by region. Progressive intentions on the part of editors were constrained by the publishers, by the advertisers, and by perceptions of what the audience would accept. Some editors and women who wrote for the magazines, though they themselves might live lives as successful career women very different from the cozy heterosexual domesticity promoted in their pages, let no hint of alternatives slip into the copy they produced. Nevertheless, as the examples quoted here suggest, women's magazines at least occasionally worked to validate women's sexuality.

So were the sex lives of British women totally transformed between 1920 and 1960? The historical record is inconclusive on this point. Numerous women wrote to Stopes expressing their gratitude for her teaching (Hall, "Science"). Hutton mentioned in her 1960 autobiography, *Memories of a Doctor in War and Peace*, the letters she had received thanking her for her work (217), but these letters do not survive. Nor do those known to have been sent to Wright (Evans 154) or those that may be assumed to have been received by Malleson. Presumably some women were achieving orgasms within marriage, given the persistent mentions within birth-control literature of the deluded belief that "holding back" from climax would act as a contraceptive. On the other hand, a good deal of evidence suggests that for many women sex within marriage was still experienced as a duty to be endured rather than a pleasure to be promoted.

Sexual ignorance continued to be extremely pervasive. Well into the interwar years, women of all classes reported marrying "with the vaguest idea" as to what conjugal relations consisted of, and even going into labor without knowing how the baby would emerge, until the sudden realization that "the child would have to come out of there" (Porter and Hall 251, Leap and Hunter 78–79). Mothers seldom did more than provide obscure warnings to their daughters, doubtless out of ignorance and shyness rather than malice. Class and educational level made little difference to this general fog. Rural Lincolnshire women in the 1930s reported brides running home after the wedding night because husbands had "tried to do something terrible," or asking the vicar to "unwed" them because of a husband's attempt at "something very rude" (Sutton 2, 90–93, 59). But educated women also experienced problems: a former teacher, aged thirty, wrote to Stopes that she had found her "first few days of married life a nightmare," and the daughter of a scientist was admittedly ignorant on "matters of sex and birth" (Porter and Hall 253).

Alternative models of female sexuality doubtless also continued to play a part. Conventional stereotypes of women as potential victims of lustful males persisted in media representations and in much didactic literature, in particular that aimed at young girls. Sexual pedagogy in schools and youth organizations (such as there was) concentrated on guarding girls against danger rather than alerting them to potential pleasure. Sexual danger was not even necessarily associated with erotic pleasure: recurrent "white slavery" scares implied that any young woman attending a place of amusement—cinema, restaurant, department store—might find herself drugged and shanghaied into a foreign brothel. This sensationalist discourse was much deprecated by serious workers in moral hygiene bodies such as the National Vigilance Association (Porter and Hall 264).

Even other works in the marital advice genre, such as the popular *Ideal Marriage* (trans. 1926) by Dutch gynecologist Theodoor van de Velde, tended to reassert male authority even while preaching the necessity of female satisfaction in the marriage bed, by situating the husband as educator and arouser of his innocent bride. A recurrent image was that of the violinist evoking music from a passive violin. Even though these works often positioned the male as lacking in expertise (indeed, as an ape producing cacophony rather than harmonious sounds from his instrument) and needing instruction, he was still assigned this responsible role (Hall, *Anxieties* 63–88).

"He's a good husband, he doesn't bother me often"

Between 1920 and 1960, there were a number of attempts at surveying women's actual experiences of the sexual side of married life. Many such studies found considerable dissatisfaction among female respondents. A rather impressionistic picture was put forward by Margaret Leonora Eyles. One of the more advanced "agony aunts" of the 1920s, whose column appeared in *Modern Woman*, Eyles wrote *The Woman in the Little House* (1922) on the basis of having lived in a working-class neighborhood. In this work she reports that several women told her, "I should'nt [*sic*] mind married life so much if it wasn't for bedtime"; one was "blooming glad the old Kayser went potty" and brought about the departure of husbands for the Front (129–32). Similarly, in her memoirs, *Still More Commonplace*, Mary Stocks reminisced about "the phrase familiar to early birth-control clinic workers: 'He's a good husband, he only troubles me once a week'" (19). And in *More About the Sex Factor in Marriage*, Wright described the kind of woman who, asked about the happiness or otherwise of her sex life, "looks quite blank and says nothing. The question is repeated. She realises that *something* is meant and asks 'Why, doctor? *What is there to enjoy*?'" (14).

This attitude was also found by Eliot Slater and Moya Woodside, in *Patterns of Marriage: A Study of Marriage Relationships in the Urban Working Classes*, based on research undertaken during the Second World War and published in 1951. They reported, "Husbands are valued in an inverse relation to sexuality: 'he's very good, he doesn't bother me much.'" The women whom they interviewed saw sex as a duty to be undertaken with "passive endurance," not an activity to be enjoyed. "A lucky minority" found real pleasure, but such experiences were far from routine (165–76). Comments made to the Mass Observation "Little Kinsey" survey of 1949[1] included "if I'd known what it was like before I got married I would never have married"; "sex isn't very nice. . . . Yes, it can be very harmful, it can ruin a woman's insides"; "if only he had made love to me instead of using me like a chamber pot" (qtd. Stanley 3).

By the 1950s the desirability of female sexual satisfaction in marriage was an increasingly widespread concept. But it is still unclear to what extent this concept mapped onto the reality of women's own experiences. Slater and Woodside commented that the younger women in their survey had greater expectations of pleasure, but it was not clear whether these expectations were fulfilled. Language itself was a problem, as difficulties in communication between investigators and the women complicated interpretation: "it was difficult to word the appropriate question" about orgasm, and "many who said 'yes' sounded unconvincing" (168–69). Two surveys of the 1950s, Geoffrey Gorer's *Exploring English Character* (1955) and Eustace Chesser's *The Sexual, Marital, and Family Relationships of the Englishwoman* (1956), suggested that sex continued to be a problem area. In Gorer's survey, sex came relatively low on the list of factors making for happiness within marriage (137–40). As with Slater and Woodside's couples, wives in Gorer's sample were wont to complain about sexual difficulties, using terms such as "selfish and demanding" and "takes me for granted" (133); conversely, others gave high marks to "thoughtfulness," which possibly had a sexual subtext (125–32).

In Chesser's massive survey, a substantial majority of women subjectively evaluated their marriages as "happy" or "very happy." But his analysis of sexual satisfaction within marriage was rendered problematic, just as Slater and Woodside's had been, because "Many women have difficulty in describing their sexual sensations, in defining the nature of their sexual pleasures, and are confused as to the definition of their genital experiences." Women had some notion that intercourse could be pleasurable; in fact, according to Chesser, many "feel ashamed to admit they do *not* experience orgasm" (421–24). The increased importance given to mutual sexual gratification in marriage brought about its own problems and pressures: the non- or infrequently orgasmic women in Chesser's survey not only had the grudges of a previous generation about male sexual demands, but also felt that their husbands were depriving them of something that was coming to be seen as a "right" (440, 448).

The disjunction between the attitudes uncovered by these surveys and the kind of marital pleasure that Stopes and her successors advocated might be attributed to class distinctions. The writers of the manuals were educated professional women and thus securely in the upper reaches of the middle classes. The subjects of the earlier surveys were all women of the working classes: the lower classes were habituated to various degrees of surveillance and intrusiveness from their "betters," so that it was doubtless considered more acceptable to quiz them about their sex lives than to endeavor to interrogate the middle classes on such a delicate topic. Thus, for instance, the 1937 Joint Council on Midwifery survey on induced abortion dealt with self-induced or backstreet abortion and made no attempt to investigate the widespread existence of abortions being performed for hefty fees by medically qualified practitioners (Thomas and Williams). And in a survey on birth-control usage undertaken by the Royal College of Obstetricians and Gynaecologists for the Royal Commission on Population in 1945, married women in general hospital wards were interviewed about their contraceptive practices, but problems were seen to exist in extending the survey to the more well-to-do women who patronized private hospitals and nursing homes (Porter and Hall 196). Even so, the Mass Observation investigators did their best to gather information from a range of classes and age groups, and—perhaps enabled by the growing prevalence of governmental, sociological, and market research surveys after the war, and indeed, by changes in attitudes resulting from war conditions—the Gorer and Chesser surveys seem to have been drawn from a wider social basis. Arguably, female sexual discontent was widespread.

"Our earlier optimism . . . was ill-founded"

This depressing situation stimulated Wright to expand upon her earlier work in *More About the Sex Factor within Marriage*. Conceding that "a certain amount of improvement has resulted" from the wider dissemination of basic information, she nonetheless considered that "our earlier optimism," the assumption that simply alleviating ignorance would do the trick, "was ill-founded" (10). From her experience in family planning clinics—and women who had got themselves to a birth-control clinic were presumably not the most cowed, apathetic, and downtrodden of their sex but had taken a degree of control over their lives—Wright claimed that "fifty out of every hundred wives still go through their years of married life without discovering that physical satisfaction can, and should, be as real and vivid for them as it is for their husbands" (11). Malleson suggested the slightly less sweeping figure of "perhaps a quarter or a third" (13–14).

Wright delineated three groups of women whose "attitude of mind" predetermined sexual failure, not only women who "expected nothing" or expected "something, but too little," but also increasingly "women who expect too much." Lacking "direct information about sex and marriage," and with misleading ideas gleaned from romances, films, and popular music, these were embittered to discover that real life was rather different (*More* 11–14). This suggestion is corroborated by a letter to the "Family Doctor" in *Lucky Star* in May 1941: "I am disappointed over married life, as I used to hear such romantic stories about the thrill of real love. I get nothing of the kind, in fact I'm getting to turn against this side of married life altogether" ("Ask" 10). A further group, according to Wright, experienced "mild physical pleasure," but had only discovered through reading works of sexual advice that they "should be capable of reaching a definite climax" (15).

Basing these conclusions on "hundreds of talks" with her patients, as well as on correspondence, Wright saw three reasons that even carefully instructed women might be unable to overcome "sexual failure": inability to differentiate "sexual response in the erogenous zones" from actual orgasm; "lack of understanding of the unique role played by the clitoris";

and assumptions, based on the male response pattern, about what a woman was supposed to feel in penetrative intercourse. The "penis-vagina fixation . . . as far as orgasm is concerned" was shared by men. The clitoris was ignored and the vagina imagined to be the "region of primary female orgasm" (44–49). This mindset surely owed less to even the most popularized and bastardized versions of Freudianism than to male-oriented "common-sense" assumptions about how sex was supposed to "work."

Why was there such a difference between the woman-focused sexual agenda that had been promoted since the end of the First World War and women's continuing negative experience of sexuality? Despite the reformers' efforts, in practical terms sexual ignorance among both sexes was still rife. While the rhetoric of mutual pleasure in marriage became increasingly common, in its most prevalent forms—for example, in women's magazines—it was, for reasons of censorship, likely to be vague about the means to the end (though books might be recommended or helpful pamphlets sent to individual inquirers). Sex education in schools was sporadic, since it was not a mandatory part of the curriculum and was therefore subject to the whims and prejudices of local education authorities and individual head teachers. If it took place at all—and according to Audrey Whiting's admittedly unreliable statistics in 1955, in five out of six schools it did not—it was often minimal. Although girls were at least likely to get slightly more attention in this area than boys, female sex education usually dealt with menstrual hygiene (indeed, as Danda Humphreys notes, representatives of sanitary protection firms were significant providers of such instruction, coming into schools with attractive presentational materials and free samples; see also Michelle Martin, chapter 7 in the present collection) and with saving oneself for marriage and motherhood, not with pleasure. Literature and other media were increasingly depicting female sexual desire, but gratification tended to be portrayed as an overwhelming force of nature rather than a learned response. Orgasm was something that happened, or that might be "given" to the woman by a man, but it was essentially out of a woman's own control, though any failure to experience it was blamed on her "frigidity" and resistant attitude.

Women's continuing lack of enthusiasm about sex is understandable. Birth control was difficult to obtain, even for married women, and not always reliable. Although family spacing was becoming much more common, it was achieved predominantly through the male methods of contraception, condoms, or withdrawal.[2] While in situations of trust and good understanding between spouses this way of assigning responsibility may have been a positive experience, in other cases it may well have been considered as controlling. That the man was in charge of contraception might also have reflected a continuing belief that sex was designed primarily for male pleasure, so that the bother and expense of any precautions should fall on him.

Conversely, sexual exploration and experimentation by women were not encouraged: premarital sex, though increasing in frequency, was stigmatized and unmarried pregnancy penalized. Even "petting," or nonpenetrational sexual activity, was usually condemned, either as likely to be the beginning of the slippery slope to premarital intercourse or because to create a "condition of frequent tension can easily upset the relation between the partners and perhaps their future attitude towards sex" (Barnes 137). Far from being considered a useful means of increasing female knowledge, petting was sometimes stigmatized even by woman writers as a form of "thus far and no further" love-making that could have a deleterious impact on subsequent marital adjustment, especially for women, who would find themselves unable to overcome immediately "the self-restraint imposed by years of practice" (Macauley 55).

The Male in the Head

Many of these factors have now changed: birth control and abortion are more reliable and more readily obtainable, sex is no longer corralled into marriage, and girls are supposed to "wanna have fun." But recent studies suggest that changing the model of sexual interaction between a man and a woman from one of penetrative sex, initiated by his arousal and concluding with his orgasm, has not proved easy.

Shere Hite's 1976 initial report on female sexuality, though criticized for what was perceived as faulty methodology—as if there has ever been a sex survey that has fulfilled all the criteria of exemplary methodology—only substantiated the comments of earlier writers; in the 2000 report she specifically pays tribute to Wright's advocacy of the need for women to familiarize themselves with their genitalia (198–200). In 1976, Hite underscored the message of these writers and the findings of the studies already cited on the inability of the accepted model of intercourse to provide gratification for the majority of women. And even in her recent expanded and updated version of this report, Hite emphasizes that as a general rule penetration continues to be defined as what is "really" sex, and that there is a "great distance" between women's own sexual feelings and experience of orgasm, and the public images of women being sexual that still embody male fantasies (12). The kinds of issues that appear in Hite's report have recently been explored by British feminist researchers working on young women and the promotion of sexual health, in the context of anxieties over teenage pregnancy and sexually transmitted diseases.

For example, Sue Lees, who has published extensively on young women in contemporary society and issues of women and sexual crime, argues that to speak of a woman's reputation still invokes her sexual conduct, and that this understanding remains "central to the way she is judged and seen both in everyday life and by the courts and welfare and law enforcement agencies" (17). Her research on adolescent girls found that the fear of being defined as a "slag" (a British term equivalent to "slut" in North American usage), whether by boys or by other girls, is a potent force for control over girls' sexual self-presentation and behavior. The only defense is to have a steady boyfriend and deploy the discourse of love to justify sexual activity (24–25). It is "somehow wrong and nasty for a girl to invite sexual activity but natural for the boy to be after it" (8). Moreover, within sex education, girls are not presented as desiring subjects or initiators but "as potential victims," having to defend themselves against "disease, pregnancy and 'being used'" (51).

Similar conclusions were reached by the Women, Risk and AIDS Project based on interviews with young women in London and Manchester; the Project's findings have been published in *The Male in the Head: Young People, Heterosexuality and Power* (1998). The authors comment in the introduction on the continuing lack of sexual agency among young women:

> We were struck by the . . . discrepancies between expectations and experience; between intention and practice; between different discourses of femininity. Generally the young women were concerned about sexual health, yet fears of sexually transmitted diseases paled in the light of their fears of pregnancy and the loss of sexual reputation. [Although they] expressed powerful intentions to practice safer sex. . . . they repeatedly failed to fulfill these intentions. (Holland et al. 5)

The authors noted "connections between an absence of female desire and the disembodied femininity that emerged from the interviews" (8). "Unfettered female desire" was almost inconceivable: "The actively desiring woman, the woman luxuriating in extravagant sexuality, seeking out men for her own sexual pleasure is a wholly negative sexual subject, if not a dangerous force." Even girls conscious of sexual desires were under considerable pressures to mute these desires and focus them into stable relationships (174–75).

Exceptional Women

In view of the continued importance of social and psychological factors inhibiting the achievement of female satisfaction, it is worth returning to the pioneer feminist writers of sexual advice in order to note the major differences between these women and the vast majority of women in Britain who were their contemporaries. Stopes, Hutton, Wright, and Malleson were highly educated professionals who had tended to marry later than average. Although Malleson, the youngest, had married while still a medical student, Wright was in her thirtieth year and Stopes and Hutton well over thirty at the time of their weddings. They had not drifted into matrimony but had had to make a definite decision to have both husband and career.

Age would seem to be a factor in women feeling more confident about asserting their sexual desires (and indeed, having a fuller knowledge of them): of Hite's sample of women in the United Kingdom, Australia, and New Zealand who were identified as having predominantly positive attitudes toward their own sexuality, nearly two-thirds were over thirty (655). The connection between maturity and sexual contentment might also underlie the well-known finding of the British Sexual Attitudes survey that both men and women found that the sex was better in long-term relationships (Wellings et al. 267–70). This finding suggests that the passage of time may bring about personal confidence and abilities for self-assertion, which might moderate the particular inequalities and lack of sexual agency felt by the young women whom Lees and the Women, Risk and AIDS Project studied.

Even given the educational and career advantages that the early pioneers enjoyed, the collision between high expectations and negative experiences had a generative role in the evolution of their theories. The most notable example of this was of course Stopes, who claimed in the preface to *Married Love* that she had made the knowledge gained in the process of discovering what was wrong with her first marriage available for the common good (xiii). Similarly, though less acutely, the difficulties that their mutual inexperience posed to Wright and her husband had a major impact on Wright's later work, and certain passages of Hutton's *The Hygiene of Marriage* suggest a subtext about the challenges of sexual adjustment in a marriage made relatively late in both partners' lives. But all these women were in a position both to encounter the idea that women should enjoy sexual pleasure, and to discover how this pleasure might be brought about.

In their own lives these women did not slavishly adhere to conventional models of married life. Stopes not only achieved the annulment of her first marriage by making a compelling case that her husband had failed to consummate the union; she also remarried, and after some years and the birth of a son, obtained from her second husband carte blanche to seek sexual satisfaction elsewhere. Hutton and her husband had a commuting marriage enabling them both to pursue their careers. Wright had several long-standing extramarital relationships. Malleson married a divorced man whom she herself later divorced, at a time when divorce was a good deal less common and subject to social stigma outside progressive circles. They were women of passionate confidence in their own opinions, which they could wield with almost hypnotic force to convince others. They were undoubtedly exceptional women who had high expectations of life and were used to getting what they wanted.

Nevertheless, in their published writings they were extremely careful to make it clear that they aimed at strengthening the institution of conventional heterosexual marriage, a claim that was probably the only way they could get their message to the public. They thus tended to collude with existing gender stereotypes in ways that undercut the radicalism of what they were advocating. In *Married Love*, Stopes depicted women as alluring nymphs, "always escaping" (70). In correspondence she was often more explicit and forthright than she felt she could be in the public arena, recommending abortion in certain specific cases and advocating masturbation as a sensible form of sexual release for the mature unmarried career woman (Hall, "Correspondents"). Only in her final published work, *Sex and Society: A New Code of Sexual Behaviour* (1968), did Wright articulate her strong convictions about the desirability of premarital experience and extramarital relationships (63). This was following, rather than making, changing social mores. Prior to the late 1960s it was only a radical marginal voice, such as that of feminist sex reformer Browne, that could openly articulate unweaving the apparently indissoluble trinity of love, sex, and maternity and posit a utopia that might include temporary and experimental liaisons, women with lovers but no children, women with children but no partners, lesbian partnerships and even lesbian mothers, and bisexual relationships, as well as freely chosen celibacy (Hall, "Woman").

Knowledge: Necessary But Not Sufficient

The authors of *The Male in the Head* point out the disjunction between "empowerment at the level of ideas" and "empowerment at the level of practical efficacy," or the ability to act on intentions (9). This is a helpful lens through which to look back at the dislocation between the pioneering writings of women such as Stopes and Wright and the persistent negative reports by women about their experiences of sex. If empowerment at the level of ideas is now much more widespread—for example, the young women in the Women, Risk and AIDS Project study at least knew about contraception and safe sex, even if they were unable to apply these ideas with consistent success—there are continuing problems for women in successfully negotiating within a sphere still focused on male attitudes and expectations.

To be sure, society no longer assumes that women will enter marriage as virgins, or at least only having had sex with their future husbands. In order to obtain a divorce they do not have to prove (as Stopes would have had to) both adultery and another matrimonial offense. Many women cohabit rather than marry. Birth control is more reliable and more widely available, as is abortion, though sexually transmitted diseases are still with us. Yet sexual reputation and the fears of losing it are still powerful constraints, affecting not merely social relationships but also interactions with judicial and welfare systems. If few women in Britain in the year 2000 could ask, "Doctor, what is there to enjoy?", they still lack an effective language to talk about sex when it is actually happening.

How much more deeply rooted must have been the factors militating against their sexual pleasure when most women were far more economically dependent upon men, constrained by the universal acceptance of marriage as the marker of social respectability, and even if they were aware of the possibility of controlling their fertility, either lacking the facilities to do so or finding them unreliable. But those who found themselves struggling against "the whole social order" for "their most precious personal right" of sexual pleasure, as Browne wrote in 1923, at least had a sense of what they were fighting for and what they were fighting against. This idea no longer has the same vivid clarity and is, indeed, positively obfuscated by media representations of modern women as "having it all."

The pioneering writers of feminist sex advice were optimistic in their belief that lifting the clouds of ignorance would immediately and vastly improve women's sex lives. While, as their own life stories reveal, knowledge of the possibility of female sexual satisfaction and the means to achieve it is an essential factor in reaching the goal they hoped for, there were other, less easily remediable, elements ranged against them.

Notes

1. Mass Observation was an organization set up by a diverse group in Britain in the 1930s with the aim of undertaking "an anthropology of ourselves" by mingling with the public and recording their comments and opinions, soliciting the opinions of people in the street, and asking a panel of volunteers to record their own observations. The "Little Kinsey" survey, unlike its eponym, did not seek to investigate sexual behavior but only attitudes, and did so by approaching individuals in the street, as well as submitting the questionnaire to panel members. For a fuller account, see Stanley.
2. Condoms had become much cheaper, pleasanter to use because thinner, and more reliable in the 1930s. The latex process was a significant technological advance, but the National Birth Control (later Family Planning) Association's issue of an "Approved List" of contraceptives led to a great improvement of standards generally in this previously unregulated industry. Arguably, their contribution to a greater reliability in "over-the-counter" methods had much more impact on declining family size than their clinics, which were, in relation to the population, few and widely scattered, and even where they existed attended by a relatively small proportion of fertile women.

Works Cited

Allister, Ray. "What's Worrying You?" *Modern Woman and Modern Home* (January 1945): 83.

"Ask the Family Doctor." *Lucky Star* 297 (10 May 1941): 13; 300 (31 May 1941): 10.

Barnes, Kenneth C. *He and She*. London: Penguin, 1962.

Bland, Lucy. "The Shock of the *Freewoman* Journal: Feminists Speaking on Heterosexuality in Early Twentieth-Century England." *Sexual Cultures: Communities, Values and Intimacy*. Ed. Jeffrey Weeks and Janet Holland. London: Macmillan, 1996. 75–96.

Browne, Anthony. "Women of the UK Feel Sex Revolution's Benefits." *The Observer* (21 May 2000): 7.

Browne, F. W. Stella. "Studies in Feminine Inversion." *Journal of Sexology and Psychoanalysis* 1 (1923): 51–58.

Chesser, Eustace, et al. *The Sexual, Marital and Family Relationships of the English Woman*. London: Hutchinson's Medical, 1956.

Evans, Barbara. *Freedom to Choose: The Life and Work of Dr. Helena Wright, Pioneer of Contraception*. London: Bodley Head, 1984.

Eyles, M. Leonora. *The Woman in the Little House*. London: Grant Richards, 1922.

Gorer, Geoffrey. *Exploring English Character*. New York: Criterion, 1955.

Hall, Lesley A. *Hidden Anxieties: Male Sexuality, 1900–1950*. Oxford: Polity, 1991.

——. "'I have never met the normal woman': Stella Browne and the Politics of Womanhood." *Women's History Review* 6 (1997): 157–82.

——. "Marie Stopes and Her Correspondents: Personalising Population Decline in an Era of Demographic Change." *Marie Stopes, Eugenics and the English Birth*

Control Movement: Proceedings of a Conference Organised by the Galton Institute, London, 1996. Ed. Robert A. Peel. London: Galton Institute, 1997. 27–48.

——. "The Next Generation: Stella Browne, the New Woman as Freewoman." *The New Woman in Fact and Fiction*. Ed. Angelique Richardson and Chris Willis. London: Macmillan, 2001. 224–38.

——. "Uniting Science and Sensibility: Marie Stopes and the Narratives of Marriage in the 1920s." *Rediscovering Forgotten Radicals: British Women Writers 1889–1939*. Ed. Angela Ingram and Daphne Patai. Chapel Hill: University of North Carolina Press, 1993. 118–36.

"Heart to Heart." *Woman's World* 2056 (17 February 1945): 11.

Hite, Shere. *The Hite Report: A Nationwide Study on Female Sexuality*. New York: Macmillan, 1976.

——. *The New Hite Report: The Revolutionary Report on Female Sexuality Updated*. London: Hamlyn, 2000.

Holland, Janet, et al. *The Male in the Head: Young People, Heterosexuality and Power*. London: Tufnell, 1998.

Humphreys, Danda. "Careers Extraordinary: One over the Eight of a Team of Travelling Teachers." *Nursing Times* 68 (1972): 578–80.

Hutton, Isabel Emslie. *The Hygiene of Marriage*. 4th ed. London: Heinemann Medical, 1933.

——. *Memories of a Doctor in War and Peace*. London: Heinemann, 1960.

Leap, Nicky, and Billie Hunter. *The Midwife's Tale: An Oral History from Handywoman to Professional Midwife*. London: Scarlet, 1993.

Lees, Sue. *Ruling Passions: Sexual Violence, Reputation and the Law*. Buckingham: Open University Press, 1997.

Macaulay, Mary. *The Art of Marriage*. London: Penguin, 1957.

[Malleson, Joan.] *Any Wife or Any Husband: A Book for Couples Who Have Met Sexual Difficulties and for Doctors*. London: Heinemann Medical, 1950.

Mechling, Jay. "Advice to Historians on Advice to Mothers." *Journal of Social History* 9 (1975–76): 44–57.

Porter, Roy, and Lesley Hall. *The Facts of Life: The Creation of Sexual Knowledge in Britain, 1650–1950*. London and New Haven: Yale University Press, 1995.

Slater, Eliot, and Moya Woodside. *Patterns of Marriage: A Study of Marriage Relationships in the Urban Working Classes*. London: Cassell, 1951.

Stanley, Liz. *Sex Surveyed 1949–1994: From Mass Observation's "Little Kinsey" to the National Survey and the Hite Reports*. London: Taylor and Francis, 1995.

Stocks, Mary. *Still More Commonplace*. London: Davies, 1973.

Stopes, Marie. *Married Love: A New Contribution to the Solution of Sex Difficulties*. London: Fifield, 1918.

Sutton, Maureen. *"We Didn't Know Aught": A Study of Sexuality, Superstition and Death in Women's Lives in Rural Lincolnshire during the 1930s, '40s and '50s*. Stamford: Watkins, 1992.

Thomas, James, and A. Susan Williams. "Women and Abortion in 1930s Britain: A Survey and Its Data." *Social History of Medicine* 11 (1998): 283-309.

True, Christine. "The Intimate Side of Marriage: Keeping Love Alive." *Mother: The Home Magazine* 2.16 (January 1938): 19.

——. "Let Christine True Help You." *Mother: The Home Magazine* 2.19 (April 1938): 118; 2.20 (May 1938): 110.

Wellings, Kaye, et al. *Sexual Behaviour in Britain: The National Survey of Sexual Attitudes and Lifestyles*. London: Penguin, 1994.

Whiting, Audrey. "They All Funk Telling Schoolgirls." *Sunday Pictorial* 2080 (30 January 1955): 13.

Wright, Helena. *More About the Sex Factor in Marriage: A Sequel to "The Sex Factor in Marriage."* 2nd ed. London: Williams and Norgate, 1954.

——. *Sex and Society: A New Code of Sexual Behaviour*. London: Allen and Unwin, 1968.

——. *The Sex Factor in Marriage: A Book for Those Who Are Or Are About to Be Married*. Intro. A. Herbert Gray. London: Williams and Norgate, 1930. 2nd ed. 1937.

Chapter 4

Molding the Man: Sex-Education Manuals for Australian Boys in the 1950s

Sharyn Pearce

As is the case elsewhere in the world, Australian sex-education manuals reflect prevailing social opinion and follow the shifts and lurches of public approbation. Prior to the 1950s, sex-instruction manuals in Australia were considered primarily suitable for marriage guidance counseling, and were mostly used by members of the clergy for discussion with young couples during their engagement period. In the public world, at least, sex was best not spoken about, and sexual information was rarely directed at young people themselves, but instead was intended to be mediated by parents, teachers, clergy, or doctors. It was during and after the Second World War that sex education for young people—"teenagers," as they began to be called at this time—came to be seen as an important part of their civic instruction, and as a prime pathway for their indoctrination into the social ethics of the time. This chapter offers some reasons for this dramatic change in thinking and analyzes some representative sex-instruction manuals of the decade, concentrating particularly upon what they had to say about and to Australian boys, and upon their underlying messages about the edification and shaping of the future man. While there were just as many sex-instruction booklets for girls, these are beyond the scope of this chapter, which chooses to focus upon the education of the undisputed "head of the household," who was expected, even before marriage, to be responsible for the "weaker sex."

In the 1950s many Australian psychologists and health workers advocated candid family discussions about sex, as they emphasized the unbalanced and unhealthy outlooks that resulted when young people received inadequate sex guidance. Experts working in these fields commonly felt that widespread sexual ignorance had increased the wartime incidence of sexually transmitted diseases. The war, and the presence of foreign servicemen stationed in Australia, had caused an enormous transformation in sexual attitudes. From 1940 onward, venereal disease in particular had become such a major problem in civilian and service life that it was frequently linked with "fifth column" activities undermining the well-being of the nation, both in the present and in the future (Davis 81–82); as one sex-education manual of 1942 characteristically put it, venereal disease was the "enemy within our gates" (Bamford, *House* 26). The reasons for its prevalence on the Australian home front are clear enough: when young men, many of them only teenagers, go off to war, the possibility of a violent end often renders meaningless the traditional doctrines and codes of behavior, and lessens previously strict moral restraints. Moreover, the 1942 arrival of the American forces meant that the long-term presence of a large, glamorous foreign army further muddied the traditional mindset by offering Australian girls, in particular, numerous opportunities for illicit sexual activity. In cities such as Brisbane, which was virtually a garrison town for the last three years of the war in the Pacific, these well-dressed and wealthy foreigners, whose courteous manner toward women contrasted strikingly with the rather casual attitude of Australian men at this time, fueled many Australian girls' desires for exciting sexual liaisons (Connors 140–63, Pace 38–40). Although many Australians still regarded the public discussion of sexual matters as indecent, the threat of a venereal epidemic was so real that one observer described it, not inaccurately, as "the worst enemy that threatens civilisation today" (Watts 93).

While escalating rates of illegitimacy, bigamy, and divorce were also responsible for inciting interest in the area of sex education, venereal disease was undeniably the prime stimulus, and the call for sex-education manuals for the young arose largely from this justifiable fear of contagion. According to widely publicized figures produced by the Australian Director-General of Health, the 1943–44 peak in venereal disease nearly tripled the prewar incidence. Data such as these caused much public alarm, promoting fears of a major public health problem (Logan 38). Although sexually transmitted diseases were routinely linked in the public mind with prostitutes and low-lives, experts declared that the majority of those who contracted and transmitted the disease were average citizens, and called for the inauguration of a Commonwealth-wide educational campaign involving the press as well as wireless broadcasts and high school lessons in an attempt to impart information in a nonjudgmental, scientific, and natural manner. Government officials hoped that such action would prevent the nation's youth from remaining either grossly ignorant or grievously misinformed.

Other factors also indirectly facilitated the promotion of sex-education manuals for the young in the 1950s. Japanese threats of invasion during the Second World War and the subsequent "stirring" of Asia, seen most ominously in the rising danger of Mao's China and the intensely nationalistic Indonesian government's desire to annex Dutch New Guinea, troubled many White Australians. The 1950s were renowned for their "populate or perish" mentality, when Australia's pro-natalist political leaders urged White Australians to have more children than ever before in order to be strong enough to counter those menaces looming above their northern boundaries. Australians were warned that they must either populate and develop their continent or accept the probability of having it taken away from them (Grimshaw et al. 271–72, Cuthbert 91). This was a time in which nationalism was still inextricably linked with the notion of racial identity, and "race suicide" (the notion that Anglo-Saxons were being outbred by Asians) still had widespread currency. The White Australia policy, which had originally been legislated by the Federal Parliament in 1901 to prevent nonwhite immigrants from settling in the country and thus precipitating racial unrest, was still tightly enforced. It was endorsed by many if not most Australians, especially given that the recent "race" war in the Pacific had been a brutal one: the appalling treatment accorded to Australian prisoners of war by the Japanese army had hardened many Australians' attitudes toward their wartime enemies, and by extension all Asians.

In the first postwar decade the Australian government sponsored the arrival of one and a half million immigrants from the United Kingdom and Europe, but allowed few Asians into the country on a permanent basis—and a jocular Immigration Minister could quip, "Two Wongs don't make a White," without stirring any protest. Echoing ideas prevalent in Britain at the end of the nineteenth century, some experts and national leaders felt that the youth of Australia needed to understand sex and develop sexual self-discipline because abuse of the sex instinct was not only wrong but might also affect future generations (Bamford, *What* 26). It was widely thought that the promiscuity and sexually transmitted diseases encouraged by moral decay could lead to racial decline and extinction, with venereal disease being transferred to future families and children. Up-to-the-minute sex education would therefore not only avert the dangers of venereal disease, but also protect the White population. In promoting good breeding habits, sex-education manuals would incidentally counsel against miscegenation and consequently against regression and the possible emergence of "primitive" physical and mental traits. They would aim to encourage young Australians to remain chaste until the time came to further the good of the nation by creating wholesome and healthy children.

It is undeniable that, despite the best efforts and intentions of concerned health care experts and educators, sex manuals published in the postwar period were in the main as repressive and reactionary as the times in which they were produced. For in the 1950s Australia was remarkable for its full employment, its steady economic growth, and its massive expansion of wealth and consumerism as a result of postwar industrialization. But the era may easily be maligned—retrospectively, at least—as a dreary period ruled by insularity and parochialism, a time of conformity and strict social divisions that worked to the detriment of minorities including suspected Communists, feminists, gays, immigrants, and Aborigines, all of whom were branded subversive elements. Although it was a time of peace, plenty, and general contentment, and Australia seemed to be a golden land offering endless opportunity and affluence, a generation of educated and talented young Australians scorned its philistinism and social and intellectual conservatism, fleeing the country for what promised to be the headier delights of decadent, arty postwar Europe. After the agonies and uncertainties of the Second World War and the threat of invasion, Australia had turned with relief to the stability of home, marriage, and social conformity, and gender roles and male/female relationships became even more rigidly defined and subject to prescribed expectations than had been the case before the war. This ethos was emphasized and heartily endorsed during the social, political, and cultural mobilization of the Cold War. Finally, Australian society in the 1950s was puritanical, with sexual attraction in the main permitted only as a legitimate catalyst for marriage, procreation, and parenthood. Prudish obscenity laws and widespread government censorship meant that the list of banned books included not only D. H. Lawrence's *Lady Chatterley's Lover* and Vladimir Nabokov's *Lolita*, but also Brendan Behan's *Borstal Boy*, Sir Richard Burton's translation of the *Kama Sutra*, and even J. D. Salinger's *The Catcher in the Rye*. Despite the urging of some psychologists and health workers, the publication of sex manuals designed to offer full and frank answers to young people's questions was never likely.

Still other factors influenced the types of sex manuals available to young people. In the 1950s Australia remained geographically isolated from the rest of the world—but its situation was rapidly changing. The influence of the British Empire was beginning to wane, and although Australian loyalty to the Commonwealth was still enormously strong, such loyalty was increasingly a matter of sentiment rather than policy. Society was undergoing a reorientation that indicated the rapid changes that had taken place in postwar Australia. The psychological and economic ties that bound Australia to the old world order had been transformed due to a mixture of foreign policy and political expediency stemming initially from the fall of Singapore in 1940, when the stunned Australian government realized that it had to look to the United States and not Great Britain to safeguard national security. This recasting of national allegiances was soon replicated in the cultural world as, under the influx of the 250,000 American servicemen stationed in Australia during the war, Australians began to gravitate away from the rural, Empire-influenced ideology that had dominated their constructions of national identity in the past and move toward an urban, Americanized culture. Those American soldiers who had been stationed in Australia during the war were part of a vanguard heralding a cultural revolution; the adoption of an increasingly pseudo-American lifestyle and American-style suburban existence would become the postwar norm.

Similarly, postwar regeneration in Australia was characterized by the increasing influence and later escalation of American consumer products, particularly American mass culture and youth popular culture. The "teenage" phenomenon—the term and concept deriving from the United States—meant that Australian adolescents with more money and leisure time to devote to entertainment than ever before were being lured by the appeal of Hollywood movies, especially James Dean's *Rebel Without a Cause* and Marlon Brando's *The Wild One* (both released in Australia in 1955), as well as by the exciting new sound of rock'n'roll, which was ushered in after the American fashion by radio disc jockeys endlessly playing Top Forty hits. The youth market was booming, in Australia as well as overseas, and so Australian magazines such as the hugely popular *Australian Women's Weekly* recognized that adolescents were important arbiters of Australian popular taste by beginning a separate monthly supplement for teenagers as early as 1954. The assimilation of images of America and American cultural materialism by Australia's youth was so profound that reporters warned that Australian teenagers' acceptance of American clothing styles, dancing, food, and slang was making them misunderstood—presumably by their elders—in their own country (Stratton 95).

With its new argot, spectacular fashions, repertoire of attitudes, and iconography, rock'n'roll seemed exciting and liberating to some Australians, yet for others it was sinister and threatening. These opposite reactions consistently reflected a generational divide. For some observers, there was a mounting sense of a "youth crisis" as young people responded to the debasing sensual attractions of American mass culture. The new tribal gangs of "bodgies" and "widgies"[1] came to be associated with promiscuous premarital sexual behavior, thought to be encouraged by rock's pulsating beat and occasionally obscene lyrics, and by the suggestive body movements of popular dance (Evans 117). In short, many teenagers were perceived as troublesome and rebellious—as a "problem." It is important to note that the boom in sex manuals in Australia in the 1950s came at this time of the disintegration of the old social order, when the traditional ethic of frugality, industry, and self-denial began to give way to an ethic of consumption and self-indulgence, and when older Australians struggled to reestablish an image of rectitude and sobriety, of nuclear family values, clean living, and decent British restraint, reaffirming British–Australian conformity, high culture, and respectability on the one hand, and rejecting questionable American excess, low (popular) culture, and unacceptable social behavior on the other. Many adult observers clearly felt that teenagers were "at risk," in need of constant adult supervision and a cautious monitoring that would reinforce respectable, middle-class deportment.

In 1947 Lady Cilento, president of the Mothercraft Association of Queensland and a highly influential and vocal advocate of parental involvement in family relationships education, commented upon the extent of sexual ignorance among young people:

> I am surprised at the numbers of young men and women (though they give every appearance of sophistication) who know practically nothing of the true facts of sex. . . . They feel a definite grudge against their parents for having told them nothing about this tremendously important part of their life and for not having told them enough to avoid disaster. Our children run the risks that they undoubtedly do run through ignorance. They should know about the whole business of reproduction. (Qtd. Logan 36)

Yet the sex manuals that ensued did not by any means satisfy Cilento's demand for material that catered to young people's needs, and the reason is manifest: Australian children and teenagers were never educated after this desired fashion by the real experts in this area. At the time, sex education was still not deemed an appropriate subject to be taught in the schools or in a public forum; it was generally believed that parents were the proper people to provide sex education, and when it became apparent that few were actually doing so, this task was largely handed over to the mainstream Australian churches. For the most part, then, sex-instruction manuals were produced not by the state governments or by health care experts, but by religious organizations, and their contents were without doubt hampered not only by Christianity's traditional notions of sociosexual morality, but also by the churches' horror at the emergence of a youth-oriented popular culture that seemingly encouraged dangerous sexual license.

In the 1950s in Australia the churches were the centers of social and sporting life, and played a major role in nearly all communities. Almost everyone identified, at least nominally, with one or another Christian denomination. Until the mid-1960s, church leaders were regarded as the moral guardians of the nation, upholding the ideal of a "Christian Australia" and using their right to offer guidance from the "law of God" to regulate public and private morality (Hilliard 145). In the public sphere the churches were eager to provide leadership and pronounce judgments, confident that religious values were universal and applicable to the whole society. It was the established view that religious institutions, by undergirding moral values with divine authority, provided the only firm foundation for personal morality and social cohesion (Hilliard 135). In the 1950s the churches were worried about stemming the alarming tide of juvenile delinquency, about those rebellious youths who were attracted toward behavior that was labeled immoral, illegal, and antisocial (Stratton 141). The churches saw venereal disease, as well as any extramarital sexual activity whatsoever, as a deep-seated moral problem, and were concerned about moral corruption and a general lowering of moral standards, especially sexual morals. Because the Christian churches were concerned about the proper management of human sexuality, they established in their publications a broad legal and moral context, discussing little about sexual hygiene but instead providing a moral education that stressed sexual self-control, the primacy of the family, the sanctity of marriage, and the spiritual side of human relationships. Sex instruction became part of an educational process at the community level that included the distribution of booklets, church youth group sessions, and occasional film evenings and lectures with what Greg Logan terms "a sound Christian emphasis" (43). Above all, despite some sectarian differences that are mentioned later in this chapter, the churches' instructions were always linked with religion and with God's wise plan for people's welfare, both spiritual and temporal.

Postwar sex-education booklets provided by religious organizations were, almost without exception, rigidly gender-differentiated: boys had their own books, and girls theirs. Their intention was, however, identical, in that all sought to work for the public good by teaching young people on the verge of puberty the currently approved sexual behavior for their age group and gender, stressing moral rather than physical contexts. And thus it happened that sex-education booklets were largely used to contain sexuality in order to keep it economically useful and politically conservative, maintaining, in Foucauldian terms, a form of "governmentality" as a means of producing conformist or docile citizens, and to perpetuate the existing form of social relations, which seemed to be threatened by the alarming pace of change. Sex-instruction manuals were therefore produced by spiritual and civic authorities who were profoundly influenced by notions of moral panic, and they were much more concerned with stemming and controlling the tide of loose morals than with dispensing useful sexual information to young people.

Most sex-education manuals, whether for boys or girls, consisted of a combination of anatomical and medical information, which was usually overwhelmed by moral and ethical exhortation. Section headings in chapter 3 of Josephine Bamford's *What I Should Know and How Shall I Tell?* (*sic*) suggest to adult educators what was then the typical pattern for sexually instructive "talks and stories" for fourteen-year-olds and over:

> Biological approach is best, coupled with talks on self-control.
> Series should include Hygiene—personal and home.
> Physiology. General, and reproductive organs of both sexes.
> Fertilisation, growth and development of the ovum, and miracle of birth.
> Venereal disease. (29–30)

Most of the manuals contained brief descriptions and diagrams of the male and female reproductive anatomy, and explanations of the mechanics of conception, pregnancy, and birth, delivered in a matter-of-fact, often brisk and clinical manner. In some of these booklets, the diagrams include carefully whited-out genital areas, but these are not the norm; in the main they provide just enough anatomical explanation to blunt readers' curiosity, and also to warn them away from the distasteful and unclean sexual thoughts, feelings, or actions alluded to in the text. In some manuals the rituals of "dating behavior" (another American term adopted in Australia at this time) are explained and the dangers of losing self-control highlighted, with the key words being "self-respect" and "restraint." Such texts recommend a neo-Victorian ethos of self-discipline as a means of controlling, if not necessarily eradicating, desire. They offer a minimal acknowledgment of sexual pleasure, emotions, or feelings, as sexual activity is presented as a utilitarian function designed to enable procreation and never as an occasion for enjoyment. Impressionable readers could even feel that sex is a medical procedure, as there is little attempt at empathizing with teenagers' feelings, and moreover a distinct reluctance to titillate. (These manuals make little or no reference to orgasm, for instance, although it is at odd times delicately referred to as a "sex pleasure ordained by God.") Although in the 1950s one out of every four marriages included a pregnant bride (and, as Peter McDonald indicates, the figure was significantly higher for weddings involving teenagers [198]), contraception is never mentioned, as it was generally considered an inappropriate subject for young, unmarried persons. Instead, the perfect preventative appears to have been "pureness of living" (Bamford, *What* 27).

It is hard to avoid the conclusion that sex educators, clergy, and other professionals were reluctant to tell teenagers what they may really have wanted to know about sex, and that sex manuals were primarily, if not exclusively, aimed at preventing teenagers from engaging in any sexual activity outside marriage. Even by the mid-1950s, when the rate of venereal infection had tapered off (Logan 42), manuals continued to emphasize the dangers of sexually transmitted diseases and illegitimate pregnancies rather than the prevention of these conditions. Promiscuity continued to be painted as "the enemy of happiness," to be rapidly followed by "paying the price" in terms of venereal disease, gonorrhea, and syphilis (with symptoms often supplied in graphic or lurid detail, an emphasis on the long-term deleterious effects, and even occasional pictures of diseased genitalia). The emphasis throughout is upon indoctrination in sex prevention, and the content is mostly judgmental oversimplifications.

During this time numerous materials were published for adults and youth leaders to stress to impressionable young people the importance of holy matrimony, a state ordained by God for the procreation of children and the avoidance of fornication. For instance, the Father and Son Welfare Movement—a branch of the Young Men's Christian Association specializing in the fields of sex education, marriage, and family life—published *Let's Get Together* for leaders of youth groups, prompted by concerns about the current fashion for "going steady," and by the need to help young people through this "maddening maze of modern bewilderment" (2). This booklet attempts in a breezy way to warn of the dangers of premarital sex and emphasizes the need to work out strategies for preventing "going all the way." It also attempts teen slang, introducing a veneer of contemporaneity that sometimes goes awry:

> The movies tell us that
> —a boy and a girl alone together
> —some music (perhaps a motor car)
> —and maybe a sparkling night
> —wrap the lot up and then you've got it. Got what?
>
> Why!! *Love* of course.

> And beware!!
> If it comes your way there's nothing you can do to prevent it happening to you.
> And when it does—boy, you're a gone coon! (11)

Later readers are warned that the "get close to me dreamboat kind of feeling" can easily get out of hand, as necking and petting can too easily take over in a relationship, leaving other, more important, areas underdeveloped. In the end sex is described as "the means whereby a man and a woman can get together with God in this ongoing work of creation" (14). *Let's Get Together* is not typical of the repressive and restrictive instruction manuals of the time,[2] but the message never varies: sex is a dangerous impulse, and it must be controlled at all costs.

Although these documents delivered, uncompromisingly, a message of abstinence, they often conveyed a gender double standard. While many ostensibly focus on sexual information, they are in fact interested in molding the man of the future in terms of moral, rather than sexual, education. Yet again, they reflect the realities of life in the postwar era, and in particular the position of men in Australian society. Historians customarily describe the years between 1950 and 1959 as "a man's decade" (Grimshaw et al. 274), in which the Australian tradition of male solidarity was further reinforced by the wartime experiences of soldiering and imprisonment. As many visitors to the country commented, men's and women's cultures were strikingly separate at this time, and Cyril Pearl's satirical 1959 look at Australian society, *So You Want to Be an Australian*, depicts manhood as consisting of drinking beer, gambling, taking it easy, and ignoring women:

> An American psychologist, Mrs Graham Bell, said she wondered how Australian men and women ever got together long enough to get married.
>
> But it is not true that Ordinary Australians fail to recognise the value of women. Any man will tell you they are indispensable for packing picnic baskets, and for keeping other women company while you are drinking with their husbands. (35)

Nevertheless, Australian men were in the process of changing, moving away from the idealized frontier image of the Australian as sunbronzed bushman or lifesaver, replete with the so-called national virtues of lusty self-reliance and mateship, as well as an inbred distrust of pretension, snobbery, and authority. In this new postwar world many men were beginning to rethink their notions about gender roles—partly as a result of the American wartime "invasion," and also as a consequence of the media promotion of a new man molded by American-style modernity and sex appeal. The Australian "type" created in the nineteenth century, the rough and ready adapter to his environment who had been elevated to the status of a national icon, had always been a romanticized figure, but by the 1950s he was probably more unrepresentative than ever. Sex manuals, however, pay scant attention to these changing sociocultural norms and patterns, just as they ignore Australia's changing demographics. Concepts such as class, race, and ethnicity are never mentioned, and it is taken for granted, despite the huge postwar influx of European migrants and "displaced persons," that the teenage male reader of the 1950s is White, middle class, and Anglo-Saxon-Celtic. In this monolithic picture of manhood, an Australian Everyboy becomes a man by learning self-control, by behaving like a gentleman, and by treating girls like the ladies that they will eventually become.

Many of the sex manuals available in Australia—particularly those aimed at the educators themselves—came from Great Britain and the United States, while the homegrown varieties that were readily available were heavily influenced by overseas trends. Two representative manuals, one Catholic and one Protestant, both of which were much used at the time and were addressed specifically to teenage boys, offer useful comparisons concerning the fashioning of the mid-century adolescent male. Patrick F. Dorian's 1959 *Instructions for Boys* (aged twelve to sixteen) links sex education with character education, and, like its companion volume, *Instructions for Girls* by Sister Mary Winefride (1962), is typical of the sex-education literature of the postwar period. Because the Catholic Church taught that human sexual desire was not merely a biological function as with animals, but a spiritual and holy act of richness, beauty, and character, we find here little about anatomy, no scientific terms, and no examples from plants and animals. In this regard, as in all others, Dorian's instructions are rigidly orthodox (the title page reveals that it was sanctioned by the Catholic Archbishop of Brisbane, and also passed by a papal censor). *Instructions for Boys* is characterized by its hortatory tone, its pervading sense of sin, its constant moralizing and didacticism, and its deliberate avoidance and erasure of the urgent questions of the young. Dorian addresses the boys directly, emphasizing his role as schoolteacher: "Just as you had to be taught in childhood—how to behave and talk and read, for example—so now you have to be told about other things as you emerge from childhood into the new world of adulthood" (3). Later he voices the Church's fear of the new materialism and world order, declaring that he will "gradually unfold a fuller picture of God's design for you and your life in all its wonder and sacredness, so that, knowing the truth, you will not be lost in the maelstrom of modern paganism" (10).

But while the manual is full of religious homilies, digressions, and abjurations, it is very short on actual information. In keeping, perhaps, with the adage that repressive times see an increase in cautionary advice literature, *Instructions for Boys* has a relentlessly theological tone. Everything is sanctioned by scripture (God's plan is repeatedly referred to as "divine," "sublime," and "wonderful"), and the text features dire warnings about the dangers of sex, which is defined in the clearest possible way as a mortal sin outside marriage. Sexual intercourse is always referred to in euphemistic terms, and premarital intercourse is presented as "breaking the law," as simultaneously shameful and shameless, for "the purpose of the sex organs is of a highly sacred character" (5). Meanwhile decent boys are sternly cautioned against listening to "so-called humorous stories" that belittle the sacredness of the "marriage act": "The intimate and sacred act of union which, with God's help, is instrumental in bringing a new life into the world is an act which belongs to the married state, and of course, must not take place between those who are not married" (12). Boys are given access only to very restrictive views concerning sexuality, learning that "The supreme symbol and act of love . . . is so sacred that it is reserved and lawful only for those who have pledged their love for each other for life, as man and wife" (11). Indeed, boys are encouraged to believe that premarital intercourse flouts civil as well as religious law, and that it amounts to a punishable offense.

Alongside the constant warnings about abstention from intercourse, Dorian offers homilies designed to discourage other unhealthy behaviors. For instance, he reassures boys that nocturnal emissions and involuntary erections are normal, but explains that excessive indulgence in masturbation is harmful and shameful:

> It may occur sometimes before you are aware of it, and this is quite normal. There is nothing wrong or sinful about it, if it just happens to you. It may be sometimes uncomfortable, but you must just wait for the penis to resume its normal condition.
>
> It is wrong, however, deliberately to cause the stiffening to occur either by turning the mind to impure thoughts, or by stimulating the penis in any way to obtain pleasure from the condition. The proper time for sexual pleasure—as this pleasure is called—is in the married state. Such pleasure has been granted by God to man and woman in order that they are willing to have children; it may not be taken by any person alone or outside marriage. (9)

Perhaps prompted by the realization that a sexual relationship in marriage would be quite a few years away, and that boys would therefore have to do something about understanding and directing the great force of sex in their lives, Dorian implies that the impulses of their own bodies should be, if not feared, then certainly vigorously and vehemently denied. And so he mentions strategies for combating wet dreams, explaining, "The best advice for preventing . . . night losses is to abstain from eating or drinking a lot before going to bed, and to lie on your side rather than on your back while sleeping" (10). The booklet does not reproduce the hysteria inherent in some nineteenth-century American warnings, which predicted that lunatic asylums and insanity lay in wait for those who indulged in pernicious habits while young (Campbell 3); instead it reassures its readers that sexual feelings are normal even as it warns of the dangers of excessive indulgence, and of the resulting feelings of guilt and shame.

Such warnings are characteristic of the 1950s. On the other side of the sectarian divide, P. L. Kenny's 1957 *Guide to Virile Manhood: A Reliable Sex Education Book for Young Men*[3] was published by the Father and Son Movement (its sister organization, the Mother and Daughter Movement, produced in 1952 a companion volume, *The Guide Through Girlhood: A Reliable Book for Young Girls*, by Kenny's wife Florence), and over a quarter of a million copies were distributed to young Protestant men, parents, and Service personnel over the course of this decade. Local Parents and Citizens Associations organized film evenings in the state schools outside school hours, open to segregated groups of Grade Seven students (eleven- and twelve-year-olds in their last year of primary school); each student came with a parent, boys with their fathers and—on another night—girls with their mothers. In this manner a generation of young Australians received what proved in many cases to be their only sex instruction, sitting awkwardly with a same-sex parent who was often, if anything, even more embarrassed than they were. At these evenings, they were given Kenny's booklet, as well as others that could also be obtained from church organizations and youth groups. According to a foreword to the ninth edition by the movement's president, Kenny's booklet "freed many young men from a sense of shame which had hampered them unjustifiably and gave them a new lead to full manhood and Christian character" (n.p.).

Kenny and Dorian are motivated by similar concerns, and both endorse family life as the cornerstone of the nation's well-being. In the same way that Dorian presents married sex as a spiritual, God-given act, but premarital sex as an act in which the body becomes a foolish and thoughtless "playground of passion" (2), Kenny links extramarital sex with sexual contamination and advises any young man who has had premarital intercourse to have a medical examination immediately after the act:

> The person who enters into sexual relationships with a girl to whom he is not married runs the risk of contracting one of the venereal diseases. . . . Untreated they can cause very serious illness, mental breakdown and even death. They can also be responsible for children being brought into the world blind, deformed, and even dead. It is not possible to tell by the look of a person whether or not he or she has one of these venereal diseases. (27)

Both writers present the tricky argument that what is holy, pure, and spiritual in marriage is base and degenerate outside of marriage, compounded purely of physical lust and lack of self-control. And, of course, both assume that venereal disease occurs only outside of marriage.

While Dorian's slight booklet is hemmed in by theology and extreme caution in addressing sexual matters, Kenny's works through the problem of male sexuality in a more detailed and rather more chatty fashion. Like Dorian, Kenny acknowledges in a section entitled "Problems You May Meet" that wet dreams and involuntary erections are normal, but he treats the topic at far greater length, advising against any *deliberate* attempts to cause an erection and against looking at suggestive pictures, thinking a great deal about sex, or overmuch handling of the sexual organs (14). Kenny appears to be operating on the implicit assumption that sexual restraint is essential for both the development of the individual and the progress of society, and that youthful habits set a pattern for adult sexual attitudes. Consequently he describes masturbation as "a selfish practice because it turns a lad's thoughts and attention in onto himself, and he becomes concerned only with the relief of his own sex urge and in bringing satisfaction to himself, changing something which should be good and wholesome into an atmosphere of unworthiness and regret" (23). While Kenny and Dorian agree that such egocentricity means that the body is not a house or temple for the Lord, and that the body should never be ill-used by unclean habits of thought or deed, Kenny articulates a more specific concern that unnatural sexual selfishness can lead to difficulties after marriage, distracting the solitary and egocentric sexual pleasure seeker from the procreation of children. He explains that the sex drive has been placed in a boy's life by a Creator "to ensure the reproduction of life" (27). As such it must not be tampered with, or misused, prior to marriage.

Moreover, according to Kenny, masturbatory impulses are essentially immature, "the product of an earlier stage of one's development that ought to be left behind" (23). For him, self-mastery has moral as well as physical and physiological consequences. His stance seems to be directly linked to the nineteenth-century belief, discussed by Patricia Campbell in an American context, that a man's sexual character was both the model and the source for the rest of his character, and that the habits of restraint developed in that area eventually led to the acquisition of virtues such as hard work, piety, and noble ideals (Campbell 4). As Kenny observes, "Unless you can understand and control yourself, you will never successfully control a business, a piece of machinery or influence others" (6).

On the whole, sex-education manuals of the 1950s reinforce nineteenth-century notions that masturbation not only causes fatigue and debilitation, but also (and especially) endangers the spirit, causing not only physiological but also moral damage (Nelson 532). Moreover, they hint that it could lead to enfeeblement if practiced to excess. Because satisfaction on a self-seeking physical plane leads to feelings of guilt and shame, boys are advised to pray in a more constructive manner in order to live more positively, and to use the power of the mind over the body. Kenny advises that if young men are plagued with vivid and lustful fantasies concerning masturbation, they should "face up to it, squarely and quietly, and say 'I am a young man now. I am growing to full manhood and this practice reverts to an earlier stage of development'" (23). Consequently, the sex drive has to be rigorously controlled, and Kenny's book is not unusual in devoting much space to strategies for avoiding thinking about sex. Suggestions include praying, but in a particular fashion:

> Some young men find themselves in a good deal of distress in resorting to prayer and yet still finding that they are not able to overcome it. We would suggest that it is perhaps unwise for a lad to pray specifically asking God to help him overcome this habit of masturbation, for this tends really to impress the habit on his mind and entrench it deeper into his life. Rather we would suggest that he asks God to help him in the development of his life generally and that God's purpose be fulfilled in every area of his life. This is positive and not negative praying. (24)

Such semantic quibbles also appear to be related to Victorian ideas that physical drives could be sublimated into the higher intellectual and spiritual spheres, and thus attain the goal of sexual purity—and "purity" was undeniably the buzzword of the 1950s. Also, perhaps as a precautionary measure, Kenny lists other ways to avoid this pernicious practice, which include not wearing tight clothing and observing the rules of personal hygiene very carefully: "the value of the daily shower cannot be overlooked. Plenty of soap rubbed into the body and a brisk rub down with a towel help to set a person in fine trim for the day" (24). The Victorians saw keeping the bowels open as a way of preventing self-abuse, and even in the 1950s this anti-masturbation sex-instruction tip remains current: "A glass of cold water first thing in the morning, a balanced diet and regularity in going to the lavatory each day will serve to safeguard against constipation" (8). Above all, exercising self-discipline means keeping the sex organs clean, a practice that will help to ensure that a boy is clean in word, thought, and deed.

Kenny goes further than Dorian by noting the damaging effects of suggestive reading material, and he cites an instance of a sixteen-year-old boy so inflamed by erotica that he committed a sexual attack upon his young sister. His reading had filled his mind "until it had so dominated his thinking, warped his outlook, that it drove him to that very tragic offence" (30). Such fears about loss of control also inform Kenny's use of personal reminiscences when he compares the sex drive to a fine horse called Carbine:

> I remember with what pride I mounted Carbine and how he pranced and champed at the bit to get going. Knowing that the bit was in his mouth and keeping a tight rein, away we went. How I enjoyed that afternoon's ride. Had this beautiful animal got the bit between its teeth or had I thrown the reins off his neck he would have got out of control and it is highly improbable that I should have had the privilege of writing this booklet which I fervently hope and pray will continue to influence and benefit the lives of many more thousands of Australia's youth. Like Carbine, the sex drive, or sex impulse, if handled wisely will render great service, but out of control will bring nothing but disaster. (33)

As this last extract amply indicates, both manuals emphasize that self-denial is not easy to achieve. They endorse the theory, then widely held, that men have greater sexual appetites than women, and that therefore boys are more likely than girls to transgress contemporary sexual conventions, either in masturbation or in premarital sexual intercourse. At a time when a newspaper columnist could talk about "the terrific power of the male sex urge" and argue that the average woman is unable to appreciate "to what extent unnatural restrictions can affect his mental outlook, his moods, and his health" (Webb 130), it was commonly thought that men and boys had hydraulic and aggressive sexual urges, whereas women and girls, naturally and ideally passive, were able to exercise restraint. This advice to young Australian women from a 1942 girls' sex-instruction booklet is typical: "Remember the standard you set is the standard of conduct which will be meted out to you. Deep in his heart, the boy will respect you and have a greater gratitude and love for you because of your help, sympathy and healthy friendship. . . . Girls, think twice before you have a hand in marring the life of your companion!" (Bamford, *House* 33).

In accord with its genesis in the Young Men's Christian Association, the Father and Son Welfare Movement's *Guide to Virile Manhood* promotes the notion of muscular manhood, with special emphasis upon chastity and self-restraint. It highlights the joy of building manliness through exercise and physical fitness, and it stresses such vital personality attributes as endurance, willpower, energy, courage, and self-control. Kenny's *Guide* advocates a doctrine of "a Healthy Mind in a Healthy Body, which should be the aim of every Australian youth" (5), the keystone in the attainment of a virile manhood that is to be created via vigorous effort and strenuous self-denial. And sexual self-mastery, the regulation of sexual activity in order to direct it into the "right" channels, is important in achieving the development from boyhood to virile manhood. Hence Kenny issues numerous instructions on how to keep fit, recommending lots of exercise, a healthy diet, and plenty of fresh air; using outdoor sports as a means of diverting sexual energies; and sublimating masturbatory impulses through joining youth clubs, taking up hobbies, and developing an all-round Boy Scout mentality. In his booklet Dorian also notes a definite link between puberty and the onset of a certain type of manliness, and declares that the chemical produced by the testicles and carried by the blood to all parts of the body, which is probably responsible for physical changes such as voice breaking and body hair, is the source of other changes as well, including "the development of manly qualities which becomes apparent in youths about this time. Keenness for adventure and the desire to excel at manly sports, for instance, are signs that a great change is taking place" (7). Such arguments are in keeping with the findings of a 1973 American study of international sexual behavior:

> The traditional Australian idea seems to be that the best way to solve sexual problems of the young is to avoid them. Thus, in practice every effort was made to de-emphasize sex, "sexiness," and sex play, and to fill the time of the young with studies and sports, channeling youthful Australian energy into activities such as team sports, aquatic activities, mountain climbing, adventuring in the wilds, and striving for championship level in a variety of sports. (Johnson and Belzer 237)

Apart from masturbation, the other worrisome topic in sex-instruction manuals for boys is homosexuality. The latter is commonly depicted as something not normal or balanced, a menace and an illness; hence the increased use of medical intervention, such as aversion therapy, psychotherapy, and drugs. Homosexuals are frequently confused with aggressive and predatory child molesters and pedophiles, who are depicted as engaging in the entrapment of innocent boys and young men. In the 1950s the homosexual was identified as a threat to Australia's well-being, and this homophobia was reflected in the sharp increase in the number of people charged and convicted in Australia's courts for "unnatural offenses" (Grimshaw et al. 127). Educators routinely stigmatized and pathologized homosexuals and homosexual desire, labeling homosexual practices as a sin and a crime and warning of the need to cultivate eternal vigilance against such degenerate and "dysgenic" habits. The term "sexual inversion" is dominant in descriptions of this type of sexuality. When the subject is discussed (and Dorian's text, for one, finds it unmentionable), it is, like masturbation, usually attributed to a fixation in an immature stage of development through which everyone passes, at a time when games, sports, and various other activities are enjoyed together by members of the same sex. Kenny views homosexuality as turning the sex impulse in an entirely wrong direction, creating strong but degenerate tendencies that may destroy a boy's life if allowed to go unchecked, and leading to unhappy men and women who can never know "the joy of marriage, home and family life that should be the happy and satisfying lot of all those who handle their life wisely" (25). It is not known whether Kenny had read Alfred Kinsey's *Sexual Behavior in the Human Male*, which had been published amid considerable furor in America in 1948 and which reported that one-third of American men had had a homosexual experience in their lifetimes, but at least his version is less censorious than that of Edward Griffith, whose *Road to Maturity* was much read by educators in Australia at this time. Griffith's text distinguishes between "active" and "passive" homosexuality; the "active" partner, who takes the "masculine" role, is described as "thoroughly pernicious and in need of curing," while the "passive" is "more feminine, the result of a combination of poor hormone action in early development and a series of unsatisfactory environmental and psychological influences." According to Griffith, young men should be always on the alert for homosexuals: "Be ready to spot them if they come your way and attempt to make physical advances" (186).

Both Dorian's and Kenny's sex-education booklets contain far more negatives than positives for boys. They appear to endorse English crime writer Dorothy L. Sayers's famous answer to the question "What does the Church think of sex?": "God made it necessary for the machinery of the world, and tolerates it, provided the parties are (a) married, and (b) get no pleasure out of it" (qtd. Webb 27). By far the most positive and reassuring elements in both booklets lie in the presentation of those forms of male/female relationships that transform desire into socially approved, platonic channels. And so Dorian cites a mutual attraction between members of the opposite sex as "part of God's design for the human race, just as in the physical world He has arranged the basic mutual attraction between the north and south poles of the magnet" (15). According to Dorian, boys like "pretty girls," listen to their "soft" voices, and are fascinated by their "different" ways of looking at the world. They delight in the company of a girl, "enjoy[ing] her charm, her happy laugh and gay chit-chat" (15). But the text then reverts to the norm, as what follows is a diatribe on the dangers of venereal disease and unwanted pregnancies, and on how "going all the way" leads to shame, remorse, loss of self-respect, illegitimacy, and blocked emotional maturity. Dorian finally offers two prototypes to guide a boy in his approach to and conduct toward the opposite sex, "Our Lady and his own mother," and claims that if he conducts himself with girls such that it will not offend either of these, "he will have nothing with which to reproach himself in later years" (16). Kenny also applauds conventional male/female courtship patterns, requesting boys to be "chivalrous and considerate" of girls and womenfolk—especially during their premenstrual days, when these charming and winsome creatures need special treatment—and emphasizing that "deep down in every girl's life is the need for real love and affection and security, and these cannot be found in any casual sex experience" (28).

As these two Australian sex-instruction manuals illustrate, 1950s hegemonic masculinity reinforced traditional places in society, prioritized purity (sexual and therefore racial), and promoted the status quo. Nineteenth-century ideals survived and flourished as authors sought to control and confine dangerous sex impulses. Both Dorian's and Kenny's sex manuals for boys offer convincing proof that the 1950s witnessed the continued promulgation of Victorian ideas concerning sexuality. Works such as these present the sex impulse as a powerful human instinct, necessary for the preservation of the species, and boys are warned by so-called experts, including sociologists and psychologists, that this vigorous instinct has to be curbed and directed by the needs and demands of social life. Sin and temperance are emphasized, with educators commonly using Biblical references to assert the virtues of self-control and self-denial, and concluding that, as T. Thornton Reed puts it in one text of this period, "the results of sexual sin in the spiritual sphere are defilement, degradation and desecration" (32). Ultimately very little real information is given to boys; as another contemporary booklet candidly notes, "All this type of sex-instruction misses the point. *The problem is not chiefly one of giving information, but of training the will*" (*Sex Instruction* 5). Boys molded after the fashion of these manuals will be paragons of virtue and ideal future citizens, but they will know next to nothing about how sex and sexuality may affect their personal lives. Which is, no doubt, just what writers of 1950s Australian sex-manuals intended.

Notes

1. Male "bodgies" and female "widgies" were a phenomenon of youth culture in postwar Australia. They were, in the main, teenagers of working-class backgrounds who, in this newly affluent, consumer-oriented world, were attracted to all forms of American popular culture, particularly rock'n'roll. Their preferred haunts were milk-bars, dance halls, and concert areas, and their preferred modes of transport motorbikes or secondhand American cars such as Dodges or Chevrolets. They wore "sexy" clothes—tight jeans (bodgies) and short skirts or pedal-pushers (widgies). They were often labeled by the conservative mainstream media as troublesome or rebellious, or demonized as "delinquent." For further information, see Stratton.
2. *Let's Get Together* is geared toward educators, not young people themselves, so perhaps it can afford to be somewhat more relaxed. But even though the educators are presumably young*ish* (leaders of youth groups were in most cases only a little older than the group members), the adoption of a fake teenage slang seems odd, and suggests a conflict in authorial understanding of audience.
3. The first edition of *Guide to Virile Manhood* was published in 1927. However, in the unpaginated foreword to the ninth edition (1957), the president of the Father and Son Movement notes: "In the past 30 years our experience of the subject has been increasingly extended and at the same time scientific and social ideas on sex have greatly expanded. It was, therefore, decided to rewrite the book in the light of up-to-date knowledge so that this new edition is much more than a mere revision." The table of contents also notes that it is "the completely revised and enlarged edition of the original book by P. J. L. Kenny."

Works Cited

Bamford, Josephine. *This House Not Made with Hands: Talks for Older Girls*. Melbourne: Armstrong Brothers, 1942.

———. *What I Should Know and How Shall I Tell?* Melbourne: John Bacon, 1950.

Campbell, Patricia J. *Sex Education Books for Young Adults 1892–1979*. New York: Bowker, 1979.

Connors, Libby. *Australia's Frontline: Remembering the 1939–1945 War.* St. Lucia: University of Queensland Press, 1992.
Cuthbert, Grace. "Filling the Empty Cradles." *A White Australia: Australia's Population Problem.* Ed. W. D. Borrie. Sydney: Australasian Publishers, 1947. 89–118.
Davis, Joan. "Women's Work and Women's Services in the Second World War as Presented in *Salt.*" *Hecate* 18.1 (1992): 64–87.
Dorian, Patrick F. *Instructions for Boys.* Melbourne: Polding, 1959.
Evans, Raymond. ". . . To Try to Ruin: Rock'n'Roll, Youth Culture and Law'n'Order in Brisbane, 1956–57." *The Forgotten Fifties.* Ed. John Murphy and Judith Smart. Melbourne: Melbourne University Press, 1997. 106–19.
Griffith, Edward. "Emotional Development." *The Road to Maturity.* Ed. Edward Griffith. London: Gollancz, 1944. 170–99.
Grimshaw, Patricia, Marilyn Lake, Ann McGrath, and Marion Quartly. *Creating a Nation.* Melbourne: McPhee Gribble, 1994.
Hilliard, David. "Church, Family and Sexuality in Australia in the 1950s." *The Forgotten Fifties.* Ed. John Murphy and Judith Smart. Melbourne: Melbourne University Press, 1997. 133–46.
Johnson, Warren, and Edwin Belzer. *Human Behavior and Sex Education.* Philadelphia: Lea and Febinger, 1973.
Kenny, P. L. *Guide to Virile Manhood: A Reliable Sex Education Book for Young Men.* 1927. 9th ed. Melbourne: Father and Son Movement, 1957.
Let's Get Together: Christian Attitudes to Boy/Girl Relationships. Father and Son Welfare Movement, n.p., n.d.
Logan, Greg. *Sex Education in Queensland: A History of the Debate Since 1900.* Brisbane: Educational History Unit, Queensland Department of Education, 1991.
McDonald, Peter F. *Marriage in Australia—Age at First Marriage and Proportions Marrying 1860–1971.* Australian Family Formation Project Monograph No. 2. Canberra: Australian National University Printing Service, 1972.
Nelson, Claudia. "Sex and the Single Boy: Ideals of Manliness and Sexuality in Victorian Literature for Boys." *Victorian Studies* 32.4 (Summer 1989): 525–50.
Pace, Helen. "All the Nice Girls: A Case Study in the Social Control of Women, Melbourne, 1942." *Hecate* 18.1 (1992): 38–59.
Pearl, Cyril. *So You Want to Be an Australian.* Sydney: Ure Smith, 1959.
Reed, T. Thornton. *The Christian Attitude to Sex.* Adelaide, n.p., n.d.
Sex Instruction in the Home. London: Catholic Truth Society, n.d.
Stratton, Jon. *The Young Ones: Working-Class Culture, Consumption and the Category of Youth.* Fremantle: Black Swan, 1992.
Watts, G. "Sex Education." *Australian Quarterly* 16.3 (September 1944): 89–99.
Webb, Elizabeth. *Stet.* Brisbane: W. R. Smith and Paterson, 1950.

Chapter 5
He's Gotta Have It: Teen Film as Sex Education

Kenneth Kidd

In his book *Teaching Sex: The Shaping of Adolescence in the Twentieth Century* (2000), Jeffrey P. Moran richly historicizes the project of sex education in the United States, arguing that from the beginning of the twentieth century, the concept of adolescence has been entangled with concerns about—and attempts to manage—the sexuality of youth. Across the rocky history of sex education, he goes on to observe, one other thing has remained constant, alongside the centrality of adolescence: our faith in schools as a site of instruction and intervention. Conservatives and liberals alike subscribe to an instrumentalist view of public education, to a "common-sense myth of effectiveness: if you teach it, they will learn" (226). "The sex educator's expectation that students will respond rationally to classroom knowledge," Moran adds, "is a peculiarly middle-class ideal" (222). In fact, research has shown that school programs have little impact on adolescent sexual behavior. More crucial are the examples of parents and peer groups, the values and practices of specific communities.

If Moran is right that socioeconomic and family factors loom the largest in the adolescent's sexual education, it may be presumptuous to credit narrative forms with any influence. In this chapter, I nonetheless assume that mainstream films made for teens provide distinct and often disturbing lessons in sex and romance—and that teens probably take those lessons to heart. Recent studies help explain the origins, strategies, and import of instructional sex-education films and videos, but relatively little attention has been paid to mainstream films designed to appeal to teen audiences.[1] Certainly such films are experienced by more teenagers—and much more often—than are instructional films or other classroom materials. They are viewed outside of school and often in the company of friends. They reflect and embody the dominant forms of community that Moran finds significant in the shaping of sexuality, among them the family, the peer group, and the class-specific neighborhood. While we cannot always know how they are received, these films clearly link adolescence with the custodial values of upper-middle-class society. Thus if other contributors to the present anthology, such as Sharyn Pearce and Michelle Martin, address print matter and sponsored puberty films used in "official" (school-sanctioned) sex-education classes, I will treat sex education not as an official program, but rather as a largely unexamined set of beliefs, practices, and texts that tend to endorse a narrow vision of adolescence and maturation.

Teen films offer several standard affirmations of the cultural status quo. First, and most obviously, they target boys and emphasize their heterosexual needs. In *Pretty in Pink*, Jonathan Bernstein argues that beginning in the 1980s, Hollywood designed films especially for "our friend, Mr. Dumb Horny 14 Year Old, because he was the one demographic whose patronage could be relied upon to give a film a big opening weekend" (4). The penis rarely appears in the teen film, while exposed breasts are par for the erotic course. Such a pattern attests to the usual privilege of the heterosexual male gaze. The female body is often directly on display, whereas the male body is presented more obliquely, less an object than a subject of desire.[2]

In addition, these films are generally preoccupied with Caucasian, middle-class rites of passage, turning a paternal eye toward the socialization of boys into consumer and even corporate culture. Few teen films feature girl protagonists; girls are secondary characters, friends, girlfriends, or sexual conquests for the boys.[3] The boys who take center stage are usually White, monied, and college-bound. They struggle with but capitulate to paternal expectations, and understand their first sexual encounter as "graduation," a rite of passage into bourgeois manhood. Even the ritualistic pact made among the boy characters, usually that they will all get laid by a certain date, turns out to be a business transaction from which they learn valuable lessons.

My impression is that Hollywood films made after 1980 tend to privatize adolescent sexuality, to confine it to the family home, even while they offer it up as the ultimate commodity-fetish.[4] Sexuality is no longer the business of the sex educator, but of the enterprising teenager operating in good faith—that is, not challenging the lifestyle of his parents. Those parents may be away, but their example looms large; we shouldn't be fooled by what looks like rebellious partying. We can see this refashioning of more public and institutional spaces into the ever-expansive home in films such as *Clueless* (1995), *Can't Hardly Wait* (1998), and *She's All That* (1999). Of course, the raising of the drinking age in the United States in the 1980s and other curtailments of teen freedom in the public sphere may also help explain this shift in setting.

It has become routine to note the absence of parents as a staple of the genre. For instance, Bernstein insists that the teen movie takes place "in a world entirely uninhabited by parents" (14). Bernstein is right to note the paucity of parents within such films. But I think he and others miss a crucial point: while Mom and Dad are often away, and while the teen film may ally itself thematically with the teenager against the parent, it also assumes the role of surrogate parent. We might even treat the moviehouse as a corporate extension of the home and family-style sex education. While parents may not approve of the teen film's indulgences, the system of ratings and reviewing attests to our conviction that film matters. Adults often express concern that teens are educated too powerfully by popular media; less often do they admit their complicity in that project. In any case, while adults are not often the protagonists of teen film, it's foolish to conclude that adult authority is being usurped. Sexuality in these films is never linked with social protest or critique; it's an expression of growing up, graduating, and getting a good job.

Finally, I suggest that mainstream teen films at once depart from and owe a debt to instructional film genres. To be sure, whereas sex-education films are designed specifically not to titillate viewers—designed, in fact, to make sex as boring as possible—mainstream teen films obviously delight in sexuality. Instructional films emphasize the biological nuts and bolts; mainstream films emphasize sexual boldness. And yet the mainstream teen film also borrows from its instructional forerunners, imitating their strategies if sometimes parodying them. We see this most obviously in the residual voice of the male narrator. In classic instructional films, that narrator lectures "in typical voice-of-God fashion" (Eberwein 2), whereas in films such as *Ferris Bueller's Day Off* (1986), the teen hero takes over that role, sometimes mimicking the adult but often sounding oddly like him. Another sex-education staple that survives the transition to popular film is the prostitute, a figure of sexual danger for the male subjects inside and outside the film.

In the sections that follow, I examine how post-1980 mainstream teen films serve as sex education as I have tried to redefine that term, taking as representative three engaging but rather different films from across a twenty-year period: *Risky Business* (1981), *House Party* (1990), and *American Pie* (1999).

Working Girls and Future Enterprisers: *Risky Business*

The mainstream teen sex film in the United States draws from two genres that Eric Schaefer examines in his study of exploitation films: the sex hygiene film of the 1910s and 1920s and the sexploitation film of the 1960s. The sex hygiene film is particularly important in terms of its naturalization of class stratification. Schaefer points out that the exploitation film originated as the sex hygiene film, initially made in response to widespread, largely urban anxieties about venereal disease in the prewar period. The first such film, Eugene Brieux's *Damaged Goods* (1913), featured an upper-middle-class protagonist (named George Dupont) infected with syphilis by a "street walker."[5]

Damaged Goods affirmed the popular belief that sexual disease came to the middle class from below, from sordid contact with the working class. This and most subsequent sex hygiene films—except, ironically, those made by the federal government—"espoused morality and continence as a middle-class defense against the threat posed by subordinate classes" (Schaefer 27). That threat was not merely sexuality, but an unproductive and ostensibly diseased form of sexuality. Naturally, it is the streetwalker's fault that George Dupont's vitality is compromised.

This standard scenario of the middle-class White male victimized by the cultural Other persists in mainstream cinema, including teen film. In *Risky Business*, set in Chicago, it is the prostitute Lana (played by Rebecca De Mornay) who brings danger, if also excitement, into the middle-class home, not only in her own seductive person but also in the accompanying form of her pimp, Guido. Lana is not damaged goods, exactly, but certainly she corrupts the protagonist, Joel, even as she helps ensure his future enterprise. She is not a streetwalker but a high-class call girl who transacts business in four-star hotels and private homes. This, I would argue, is an important symbolic conflation; sexual liberation translates into sexual commerce. Perhaps because she is such a good businesswoman, Lana is presented as entirely safe. Joel never worries about sexually transmitted diseases, and in fact *Risky Business*, made just before AIDS became a household word, has nothing to say about safe sex or even birth control.[6] The film updates the sex hygiene film, appealing to the same class differentiation but making clear that profit is more important than, and can even redeem, a flawed character. Lana's forgivable flaw is her ruthless business sense, not her contagious sexuality.

Risky Business also gives priority to the reproduction of the paternal estate. Tom Cruise's character, Joel Goodsen (good son; get it?), is introduced as an innocent and naïve boy with a lot to learn, but before long he's had sex with a call girl and set up a house of prostitution in his parents' mansion. Presumably this is the risky business we're watching for, although there are other possibilities; I'd venture that masculinity and paternal authority are risky business too, if also business as usual. Certainly the film is just as concerned with escorting Joel into a secure adult male selfhood of an upper-middle-class persuasion as it is with introducing him to the pleasures of the flesh. *Risky Business* is not really a movie about getting laid. Instrumental to Joel's makeover from boy to capitalist are corrupt father surrogates, chiefly the pimp Guido but also a recruiter from Princeton who enjoys this home visit more than most. Joel's parents are indeed away. Even so, Joel is not rebelling against his businessman father, but rather evolving into him, if through an unorthodox entrepreneurial venture.

Heterosexual desire, peer culture, and future enterprise are linked from the opening dream sequence of *Risky Business*, in which Joel fantasizes about a sexy neighbor, then finds himself in a classroom, watching helplessly while his peers take their college boards. He's narrating this dream to his cigar-smoking, poker-playing pals, who tease him about his virginity even as they gossip about beginning salaries of Ivy League graduates. Joel is encouraged by his friend Miles, himself Harvard-bound, to make the most of his parents' absence. Miles starts the fun by dialing up an escort service and pretending to be Joel. Here, as throughout the film, our hero is an innocent victim of the wily tricks of others. "A good time," emphasizes Miles, quoting from the ad, "in the privacy of your own home; what else could you ask for?"

But Joel didn't ask for and doesn't appreciate "Jackie," the tall, attractive African American drag queen who shows up. Indeed, he doesn't even understand the masquerade. Joel tries to turn her away; Jackie complains that she never ventures so far into the suburbs without compensation, and Joel pays her the usual fee of seventy-five dollars. But before she goes, Jackie gives Joel the phone number of another escort service and tells him to ask for Lana. "It's what *you* want," she says a little disdainfully. "It's what every white boy wants," and crucially, every White heterosexual boy. Later, a desperate Joel dials the number, and soon Lana arrives. She is thin, blonde, and generically beautiful, a striking contrast to Jackie. Joel is mesmerized. In a postmodern, almost campy sequence of scenes, they have sex throughout the house. We see them shagging on the stairs and in the easy chair, as the camera pans across various photographs of Joel as a little boy; presumably we're to contemplate his loss of innocence. Such loss is sadly clichéd in *Risky*

Business. Just imagine the transformative possibilities of a close encounter with Jackie! But this kind of encounter, of course, never happens in such films. Diverse sexuality is a joke, not a profitable venture.

In the morning, the savvy Lana sizes up the house and its contents. "Nice place you got here," she tells him. "Is it all yours? Do you know what it's worth?" When he tries to pay her for services rendered, he's stunned to discover that her fee is 300 dollars (too bad he didn't stick with Jackie). As even he doesn't have that kind of cash handy, he goes to the bank and cashes in a savings bond, a gift from his grandparents, but by the time he's back, she's left the house, having stolen a Steuben glass egg that is the treasured possession of his mother. He tracks Lana down, helps her escape from her pimp, Guido, then finds himself unable to get rid of her. She takes to cooking Joel breakfast and wearing Princeton sweatshirts. (Joel's? Dad's? We're never sure.) She also continues to take inventory. Meanwhile, her colleague Vicky has moved in, and before long, there goes the neighborhood. Guido shows up and threatens Joel, but in an oddly paternal way. "I like you," he says, putting his arm around the boy. "You seem like a smart kid. Let me tell you something. In a service economy, never, *ever*, fuck with another man's livelihood." He demands the return of his property, or else.

But Joel doesn't have much say in the matter, especially after Lana, in her devious way, causes him to dump Daddy's Porsche into Lake Michigan. Joel is desperate for repair cash. Thus begins the best little whorehouse in the suburbs. As Lana and Joel make the necessary arrangements, drumming up support among their respective friends, Joel comes to admire Lana's "shameless pursuit of immediate material gratification," as he says in a voiceover. "What a capitalist!" The opening night is a success, and not just because so many horny teenage boys come away happy, or because Joel rakes in so much cash. The recruiter from Princeton appears, and although Joel is decidedly not Ivy League material, his penchant for business is obvious. By this point in the film, Cruise is regularly sporting Ray-Bans, ever more impervious to adolescent anxiety and adult intimidation (sales of this brand of sunglasses quadrupled the next year). "We could use a guy like Joel at Princeton," concludes the recruiter in his follow-up phone call to Dad. Here the link between academic and corporate culture is made explicit.

All of this is presented tongue-in-cheek, including the poke at Ivy League privilege and hypocrisy. Even so, I don't think we're to understand the film as only amusing or ironic; *Risky Business* may make fun of venture capitalism, but it certainly doesn't challenge it. All along Joel has belonged to an extracurricular club called Future Enterprisers, which is something of a joke; students cautiously invest in stocks and report their miniscule earnings at year's end. Joel's stint as a pimp is infinitely more worldly and profitable than the club's laughably academic business ventures.

Before the film ends, in fact, Joel is forced to appreciate Guido's lesson in free/future enterprise. As retaliation for Joel's theft of "the girls," Guido steals all the furniture in the Goodsen home, just as the parents are returning to Chicago, and makes Joel buy everything back. Lana is nowhere in sight, but Vicky helps Guido complete sales. This isn't simply payback; it's a lesson in net profit, or loss. Guido recoups all the extra money that Joel made and, in fact, charges 300 dollars for the glass egg after Joel is broke, which means that financially, Joel is back where he started.

Earlier in the film, Mr. Goodsen was the less trusting of the two parents; he specifically warned Joel not to drive his Porsche and to treat his beloved hi-fi with care, saying, "If you can't use it properly, you can't use it at all." Mom, on the other hand, trusts Joel. But at the film's conclusion, the parents' roles have reversed in this respect. Dad does not discover that the Porsche was so recently underwater, and he's thrilled about Joel's admission to Princeton. Mom, however, detects a crack in her egg, one that nobody else can see, and is furious with Joel. "I trusted you," she says bitterly. Obviously *Risky Business* is having some fun with the Oedipal dramas of family; Dad is happy because Joel has identified with him, while Mom understands that she's been replaced by another object—and worse still, by a prostitute.

In any case, *Risky Business* offers its own take on *Damaged Goods* for a Reagan-era audience, distinguishing between the upwardly mobile young man and the shady prostitute. More overtly than any other teen film I've seen, *Risky Business* shows us what heterosexuality looks like under late capitalism. Sex is privatized, kept at home literally and ideologically, even as it also serves as an analogy for big business.

Hey Mr. D. J.: *House Party*

Adolescent male vulnerability-turned-triumph is a standard theme of teen sex films; we laugh as a horny, awkward teenage boy stumbles through close encounters with the opposite sex. That heterosexual stumbling often gets ritualized as dancing, and/or dancing lends that stumbling a redeeming style. *Risky Business*, of course, set a precedent with the famous scene in which, on his first night home alone, Joel strips down to his briefs and dances to Bob Seger's "Old Time Rock and Roll."

Dancing is at once fun and serious, innocent and sexy; it reveals both the grace and the vulnerability of the adolescent body.

Dancing sexualizes the young male body in a productively heterosexual fashion. Teen films at once demand and disavow homosocial, even homoerotic identification on the part of male viewers.[7] Even so, desire and identification are hard to separate in the phallic economy of the mainstream teen film. In *Risky Business*, the dance scene is potentially more unsettling than in later films, because there are no girls around, and also because Cruise is now a major sex symbol for men as well as women, thanks to this very film. Presumably, it's acceptable for male viewers to watch because he's home alone, but still, I'd be curious to know whether straight-identified male viewers find Joel's dirty dancing titillating. In any case, our cultural faith in heterosexual dancing makes same-sex dancing seem radical, if also strangely innocent, which is perhaps why I find the concluding scene of the British film *Beautiful Thing* (1996)—in which two adolescent boys declare their love by slow-dancing together in public—so sweetly satisfying.[8]

In both *House Party* and *American Pie*, dancing is a form of heterosexual seduction or courtship, but once more safely contained in the middle-class home. No one cuts footloose, or even dances in the streets, as in *Ferris Bueller's Day Off*; dancing is not dirty but good clean fun, a form not of social rebellion but of hopeful foreplay and comic exuberance. In *House Party*, written and directed by Reginald Hudlin (and produced by his brother Warrington), dancing is also an important expression of peer solidarity and the social life of adolescence. Its central character, Kid (Christopher Reid), is nerdy but just impish and adorable enough to escape ridicule. Kid and his friends Play (Christopher Martin) and Bilal (Martin Lawrence) are high school kids who just wanna party. When his parents leave town, Play throws a dance party. Kid and Play are the life of the party, performing first a mini-dance-off with girl friends Sharane and Sydney and then a boy–boy rap-off. That it's great fun to watch simultaneously reflects its apparent lack of any pedagogical agenda and heightens its ability to impose one.

For *House Party* represents at once a significant intervention in the American teen film and a standard middle-class story of coming of age. Released on the heels of Spike Lee's *School Daze* (1988) and *Do the Right Thing* (1989), and the year before John Singleton's 1991 film *Boyz N the Hood* (a tribute to the strong African American father), *House Party* features an all-black cast of characters, with the exception of two White, racist, bumbling policemen. Aside from these men, who are slightly dangerous at best, there are no menacing characters. Everyone lives peacefully in the same neighborhood, which is lower middle class but includes the projects. This is not an urban battleground, just your average neighborhood, like the white-bread environment we know from mainstream cinema. The film is not political in the way that Lee's films are, but it does give us a positive glimpse of middle-class African American life.

One way in which this glimpse is affiliated with mainstream ideals is that the boys are cute, but they are not sex symbols; predictably, sexiness is left to Sharane, the girl from the projects. Once again, motion shields the viewer from uncomfortable moments of voyeurism, and Kid is nearly always in motion. The White boy protagonists of teen films often play sports; for Kid and his boy friends, dancing is the preferred form of physical education and engagement, not just with girls, but also with each other. When Kid's not dancing, he's running—running through the neighborhood, running to escape bullies, running in circles in a jail cell. But this running is comical; here, aimless running is not, as it is elsewhere, a trope for the despair of underclass youth.

Kid would never run away from home, or feel that home was running away from him. Quite the opposite. Kid's mother is dead, but even that doesn't seem to trouble him much. He lives securely with his strict but loving father, and as in *Risky Business*, the father's values ultimately govern the son. The plot of *House Party* is a simple away-and-return narrative: Kid gets in trouble at school, gets grounded, sneaks out to the party at Play's, has fun, gets the nice girl, comes home, and gets punished. The film begins and ends with Kid at home in bed, being tended to by his father, played by Robin Harris. Pop works nights, and as the film starts, he wakes Kid up and makes breakfast for him, only to fall asleep on his son's bed. Tenderly Kid unlaces his work boots and tucks Pop in. Pop clearly loves Kid and pays attention to his life, unlike so many distant fathers in teen films. In turn, Kid loves and respects his father, and we know that his disobedience won't threaten their bond.

Pop knows that Play's parents are away; if they were home, he says, he'd consider letting Kid go to the party. But after all, it is a school night. "You'll make me a social misfit," complains Kid. "I'm trying to make you a responsible person," replies Pop. "But I'm *too* responsible," tries Kid. Pop's not persuaded. "If you mess around," he tells his son, "you'll make the same mistake I did." Presumably that mistake was getting Kid's mother pregnant; Pop envisions a more secure financial future for his son. Pop immediately apologizes and reaches out to Kid to touch his shoulder, but he stays firm. Surprisingly, when Kid sneaks out, Pop hears him but decides not to go after his son, at least not until later, when Kid hasn't returned by curfew. Pop knows exactly where Kid is; Kid has not driven Daddy's Porsche downtown or into the lake.

The sexual politics of the film likewise validate the father's perspective. Kid is interested in both Sydney and Sharane, who are fast friends. (In fact, Sydney seems jealous of Sharane's attention to boys, and throughout the film their friendship is intense, even if lesbian possibility is foreclosed.) At the party, Kid talks to Play about the girls, and Play encourages his friend to pursue Sydney, because her middle-class lifestyle creates greater opportunity for privacy: "She has a home, a house, with a rec room. You know what time it is. Go downstairs into that basement . . . sofa bed . . . bam!" Sydney's parents own several grocery stores and live in a beautiful two-story house. In other words, Kid might enjoy with Sydney not only privacy, but also a secure future through upward mobility. While it's Sharane whom Kid really wants—she is more conventionally sexy, being busty and partial to revealing outfits; Sydney is flat-chested and more conservatively dressed—she has no privacy at all in public housing, and thus little potential.

After the party, Sydney proves herself not only sexy, but also a tease: she flirts with Kid but won't kiss him after leading him on. By this point, the viewer has been cued that Sydney is jealous not because she wants Sharane to herself, as seems likely earlier, but because she wants Kid. Kid then walks her home. They kiss and say their goodbyes, and the scene suddenly shifts to her bedroom, where they are making out wildly, so wildly that we can hardly see anything. Fumbling with Sydney's bra, Kid says, "I have a question for you. Are you on the pill?" *House Party* here pays homage to the sex-education project, while acknowledging the pleasure of sex.

They stop short of intercourse, however. First, Sydney is *not* on the pill, and she doesn't have a diaphragm handy; "I don't do this sort of thing often," she explains. And Kid doesn't have a condom. "There'll be other times," reassures Sydney. "That'll be a long time from right now," he replies glumly. Just then, her parents return home. There's the usual scrambling for clothes, and Kid bolts. Later, as Kid is finally heading home, he tells Play and Bilal what never happened. They're surprised that the lack of a rubber stopped him. "I'm just not about having babies right now," he retorts. "You sound like your Pop," complains Play, linking safe sex and paternal discipline.

And in fact, *House Party* winds down with the reappearance of the father. First Pop shows up at the party, just as Kid is running out the back door. In the final sequence of the film, Pop roams the neighborhood in search of his son, as Kid and some local bullies who have been tormenting him throughout the evening get thrown in jail. These bullies provide much of the comic relief, indulging in revenge fantasy that smacks of homoerotic sadism while it simultaneously evokes the sex-hygiene discourse of much earlier films such as *Damaged Goods*. In jail, Kid has even bigger bullies to deal with, and they too want a slice of him; one particularly menacing dude tells him, "I used to run a day-care center." As they circle him lustfully, Kid again launches into song, this time an inspired rap about the dangers of sodomy that links anal sex with AIDS and ignores the issue of protection: "Speaking from the heart, from the center: the sign on the ass says 'do not enter.'" "Ask yourself, 'can it wait?' or do you wanna become an AIDS candidate?"

Finally back home, he sneaks into his room, strips down to his underwear, and slides into bed. The final shot of the movie is of Pop, looming in the doorway with eyes ablaze, snapping a belt and saying, "Your ass is mine!" *House Party* thus subordinates perverse desire to the law of the father presiding over the lower middle-class home. Pop *does* own Kid's ass, after all, and a little discipline is in order, at the hands of a real father. In the end, you can't escape Dad—but why would you try, if you've internalized his values and courted a decent girl from a decent family?[9]

Home Sweet Home: *American Pie*

In 1947, Coronet Instructional Films released *Shy Guy*, starring a young Dick York (of future *Bewitched* fame) as Phil Norton, a transfer student trying to adjust to a new high school. In fifteen action-packed minutes, we see Phil transformed from an unhappy loner into a popular guy. At the film's beginning, Phil lurks about the school, clueless and forlorn, as narrator Mike Wallace intones, "You imagine they keep watching the way you look, the way you act. They think you're different. So, you head for home. But still you can't forget that you're alone, an outsider." At home, Phil fiddles with radios in the basement. Dad, a cheery businessman, descends in his three-piece suit, and after hearing of Phil's troubles, likens his son to the oscillator the boy is trying to install in the radio, saying, "You have to fit it in with all the other parts." He has this advice: "Pick out the most popular boys in school and keep an eye on them."

Dad himself abides by this ethic of imitation. When they first moved to their new town, he tells Phil, he had his own trouble interacting with coworkers, until he learned to adopt their habits and attitudes. Not surprisingly, what works at the office also works at school. Phil follows Dad's advice, and his social life soon picks up. He attends a mixer and wows his peers with his knowledge of electronics. "They know he's *alive* now," the narrator explains. The film implies that Phil is well on his way to romance, too. In the last scene, Phil entertains his new buddies (boys and girls alike) in the basement, as a smiling Dad serves up a tray of Cokes.

According to Ken Smith, "*Shy Guy* was the most successful social guidance film of its time and was the first film for teens that pushed 'fitting in,' rather than regimented rules of etiquette, as the way to social happiness" (203). *Shy Guy* is an example of what Smith calls "mental hygiene" films, which flourished from the mid-1940s through around 1970, made for use in junior high and high school classrooms and addressing issues such as table manners, posture, dating etiquette, highway safety, juvenile delinquency, and substance abuse.[10] Some mental hygiene films target girls, such as the menstruation film *Molly Grows Up* (1953), produced by Medical Arts Productions on behalf of Personal Products Corporation.[11] Most of these films, however, address boys, and use a male narrator, the paternal voice of authority.

Coronet Studios, which produced *Shy Guy*, was run by David Smart, publisher of both *GQ* and *Esquire* (and a frustrated movie mogul), and we can certainly see continuities between these men's magazines and social hygiene films, which stressed the importance of social grooming, peer conformity, and deference to male experts and role models. At first, Dad's advice in *Shy Guy* seems counterintuitive. He tells Phil not to be proud of his individuality, but to ape his successful classmates, particularly "the other fellows." But we trust Dad, even if his advice is more assimilationist than we might expect. Mental hygiene films seem to encourage greater individuality but require deference to paternal authority.

Shy Guy seems today a quirky little film, but I find it strangely similar to the 1999 blockbuster teenpic *American Pie*, directed by Paul Weitz. This film is likewise organized around a needy and nerdy teenage boy, whose father also stresses the importance of fitting in. Like Joel and Kid before him, Jim (Jason Biggs) is a sexual innocent. And like Phil in *Shy Guy*, Jim is taken with technology—in his case, the computer, and specifically the webcam, which provides assorted voyeuristic thrills (whereas Joel toys with Dad's Porsche and Kid plays disk jockey).[12] As *American Pie* makes clear, sex education is a family affair. While the film concentrates on Jim and his peer relationships, the father–son bond is also important. Dad tries to share sexual information with Jim, buying him girlie magazines and encouraging him to find a partner. Never does Dad advise abstinence; never is he angry or even impatient with Jim's sexual shenanigans.

With *American Pie*, we return to the White middle-class neighborhood; for a film made in the late 1990s and set in southeast Michigan, it is astonishingly monoracial. We return also to the safe house of the father, with several critical differences. Jim's father, played by Eugene

Levy, is not a disciplinarian, asserts no ownership of Jim's ass. Like *House Party*, the film opens and concludes at home, situating Jim's escapades within a safely middle-class world. In the opening scene, Jim's parents catch him masturbating to porn, and Dad does his best to diffuse the situation. At the end of the film, Dad watches his son dance in his bedroom and then dances down the second-floor hallway himself, as the credits roll. Dad looks decidedly retro, with his starched white shirts and his nerdy glasses; presumably this is how Jim will look in twenty years.

Jim, like Kid, is honest and trustworthy, a model son. He is much less resourceful than Kid, and more awkward with the ladies. The film revolves around the friendship between Jim and three other boys—Kevin, Finch, and Oz—and the pact they make (you guessed it) to get laid by prom night. But while the film pays homage to *Porky's* (1981) and its raunchy ilk, it also represents a kinder, gentler sort of teen film, one that emphasizes romance, responsibility, and male vulnerability. Nearly all teen films revolve around the threat of sexual and social humiliation, but in *American Pie*, humiliation takes assorted forms and has assorted consequences. The standard geek character, for instance, brags about sexual conquests he didn't pull off and is then humiliated at the prom. Finch, one of the good guys, is also humiliated, but recovers his reputation when he scores with an older woman. That woman's son, Stifler, Finch's oppressor, is humiliated when he stumbles onto the unlikely pair of lovebirds. Oz is ditched by a college woman when he tries to put some ridiculous moves on her.

But it's Jim, of course, whose vulnerability is the source of the film's humor. His sexual urges are landing him in all sorts of odd situations. Having been told that third base is "like warm apple pie," Jim can't resist his mother's freshly baked dessert and humps the pie spread-eagled in the middle of the kitchen. Actually there are two editions of the film; in the censored version, he's gratifying himself more discretely in a corner of the kitchen. In both versions, Dad stumbles onto the scene, to Jim's dismay. Earlier, Jim is horrified when Dad shows up with his big brown bag of dirty magazines; he is also horrified when Dad discovers his stash of condoms.

When he leaves behind baked goods, Jim has an equally hard time. All along he's had a crush on Nadia, a Czech exchange student (played by an American actress), and halfway through the film, he invites her over. He leaves his room so she can change, and she quickly gets naked. She also finds Jim's girlie magazines and starts masturbating, to the astonishment of his schoolmates, who are watching courtesy of Jim's webcam (he mistakenly E-mails the site address to all his classmates).[13] This is the communal gaze, and it is coded as male within the film, as we only watch boys watch the sexual scene. Since there aren't any shots of girls watching, girl–girl eroticism is safely restricted to Nadia's perusal of pornography inside Jim's bedroom.

Jim himself has been watching at Kevin's house and rushes back home when his friends urge him to trade in voyeurism for the real McCoy. But when he tries to put the moves on Nadia, she retaliates, saying, "You've seen me. Now it's my turn to see you. Strip!" Jim does a strip-dance for her. Twice, Nadia tries to engage him sexually, and each time, Jim ejaculates prematurely. Surely this is the ultimate disgrace: to come too quickly, twice, in front not only of a gorgeous babe, but also of all of his peers watching in cyberspace.[14] Near the film's end, Jim scores at last, with a nerdy band-camp girl who turns the sexual tables on him, yelling, "What's my name? Say my name, bitch!" The next day, she's gone. "I was used!" he says. "Cool!" She's more knowledgeable about sex than Jim or any of the boys, which further emphasizes male innocence.

But the very conditions of Jim's vulnerability assure us that he's not in any serious danger. He's a normal boy with normal urges, the awkward but willing boy-toy for foreign and/or oversexed young women. Apparently it takes a big-breasted, long-legged, masturbating "Czech" beauty to make American pie. Just as *Risky Business* makes use of the streetwalker convention of the sex hygiene film, *American Pie* borrows from the sexploitation film, popular since the 1960s, which revolves around an exotic, seductive female beauty like Nadia. Sexiness, like disease, is imported into the middle-class home from the outside, or from below. Most sexploitation films took the form of nudie pictures and were often promoted as art films; some were foreign in origin, while others were made in America.[15] The sexploitation film flourished until the late 1970s, when nudity became more standard in mainstream cinema and when hardcore pornography became a more established genre (Schaefer 339).

American Pie may be an improvement upon the traditional teen get-laid picture to the extent that the girls do have some power and authority. But Jim's experiences with Nadia and then the band-camp girl suggest sexploitation alongside a more progressive representation of female sexual agency. Whereas the four boys seem to have about the same level of sexual experience, the girls and women differ greatly along these lines. Nadia, of course, seduces the hapless if quite willing Jim. Finch is seduced by Stifler's mother; they drink scotch and then have sex on a pool table, with "Mrs. Robinson" playing in the background. Jessica is a wise if jaded classmate who manipulates Finch and gives advice to the other girls. The classier, college-bound girls are depicted as more naïve and innocent but still very much in charge of their lives. Kevin's girlfriend Vicky calls the shots in their relationship, while choirgirl Heather persuades Oz to miss a lacrosse game so that he can sing a duet with her in the state choral meet. These two girls, especially, want love, not just sex. But *American Pie* is not the girls' story in any sustained way, and its reliance on "comic" female types—the sexy foreigner, the libidinous older woman—makes even clearer its investment in male innocence.

Jim and his buddies may not be headed to Princeton after graduation, like Joel in *Risky Business*, but their socioeconomic future seems promising enough. Few teen films, of course, are as shamelessly "enterprising" as *Risky Business*. But both of the two later films naturalize social class and social privilege through sexuality and coming of age. Even *House Party* winds up disavowing the girl from the projects and keeping its male protagonist safely ensconced in the middle-class world.

Conclusion

Whereas American films about adolescence are remarkably heteroconservative, with straightforward plots about nice middle-class White boys looking to "graduate," European films routinely show adolescence as a time of sexual curiosity and exploration. It is no coincidence that most films about lesbian/gay children and teenagers are produced outside the United States—among them *Du er ikke Alene* (1978), *Venner for altid* (1987), *Voor een Verloren Soldaat* (1994), *Les Roseaux sauvages* (1994), *Beautiful Thing* (1996), *Ma Vie en Rose* (1997), *Get Real* (1998), and *Edge of Seventeen* (1998).[16] My guess is that the major audience for European coming-of-age films is composed of European and American adults, and the difference in target audience may help account for what's deemed appropriate or viable on screen. Children and teens, after all, are the major consumers of teen film in the U.S.; presumably the studios assume that "Mr. Dumb Horny 14 Year Old" (to recall the label that Bernstein, perhaps unfairly, has given to the typical consumer of this type) isn't interested in same-sex love stories, for example, although one might suggest that if *The Rocky Horror Picture Show* (1975) remains such a midnight cult favorite among teens, surely there's at least *some* demand out there for alternative screenings of adolescent sexuality.

Yet if that demand exists, it is only rarely being satisfied; mainstream American films about adolescence seldom offer an alternative take on coming of age. The handful that might seem to resist the dominant sexual ideology includes *What's Eating Gilbert Grape?* (1993), *Welcome to the Dollhouse* (1995), *Kids* (1995), the road movies of Gregg Araki, perhaps *Girl, Interrupted* (1999)—but in general, there are few films that seem as countercultural as *Harold and Maude* (1978), in which a teenage boy develops a loving relationship with a much older woman. Similarly, there are very few American movies that celebrate the sexuality of adolescent girls. One recent exception is Colette Burson's *Coming Soon* (1999), about a group of upscale Manhattan girls searching for sex; it almost didn't find a distributor, and after release garnered very negative reviews.

Though not a programmatic form of sex education, mainstream teen films teach adolescents about options in love and life, steering them away from sexual and cultural heterodoxy and emphasizing the pleasure and profit of normative desire. Such films have perhaps capitalized on and compensated for the failure of sex-education programs in the schools. We should probably reserve judgment about the influence of Hollywood's products until we have better information about how they are understood by teenagers. Even so, we need a broader understanding of sex education, and closer examination of mainstream teen film, which so consistently emphasizes the sexual needs of the middle-class straight boy. For the effect of pitching teen films to a youthful audience and leaving them outside the scope of serious analysis, ironically enough, may have been to enshrine precisely the patriarchal values that adolescents are sometimes presumed to be rebelling against.

Notes

I am grateful to the editors of this volume, and to Patrick Brennan and Sid Dobrin, for their useful responses to an earlier version of this chapter.

1. I am particularly indebted to two recent books, Robert Eberwein's *Sex Ed: Film, Video, and the Framework of Desire* (1999) and Eric Schaefer's *Bold! Daring! Shocking! True! A History of Exploitation Films, 1919–1959* (1999). Other relevant studies include Kevin Brownlow's *Behind the Mask of Innocence* (1990), Lisa Cartwright's *Screening the Body: Tracing Medicine's Visual Culture* (1995), Alexandra Juhasz's *AIDS TV: Identity, Community, and Alternative Video* (1995), Cindy Patton's *Fatal Advice: How Safe-Sex Education Went Wrong* (1996), and Martin S. Pernick's *The Black Stork: Eugenics and the Death of "Defective" Babies in American Medicine and Motion Pictures Since 1915* (1996).
2. Instructional films designed for both sexes rarely show genitalia or breasts, preferring representations such as blackboard sketches, animated egg and sperm characters, and even, in the Procter & Gamble film *I Got It!* (1988), a mock-up of the female reproductive system in pancake batter. In such instructional films, social hygiene is as important as sexual hygiene; "It" in *I Got It!* refers both to menarche and to a party invitation.
3. The genre's sexism explains why Molly Ringwald, star of three popular John Hughes films, most notably *Pretty in Pink* (1986), was hailed as a breakthrough teen star and featured on the cover of *Time* magazine. But girl characters are still peripheral in teen film, even (perhaps especially) when they are as sexually assertive as the boys. In her essay on Hughes's teen films, Rochelle Mabry concludes that despite signs of progress, a conservative gender ideology marks the Hughes canon. In her view, the filmmaker "visually and thematically

effaces an individualistic teenage girl's character in favor of a feminine ideal that conforms to—and comforts—relatively conservative adult tastes."

4. In a recent essay, Wheeler Winston Dixon credits Samuel Z. Arkoff and Jim Nicholson, cofounders of American International Pictures (AIP), with recognizing in the 1950s that "what teens of that era required more than anything else were escapist visions of a teenage life without adult supervision, but with plenty of money (and the things it buys) in evidence in every luxuriously appointed sequence" (129). Clearly such materialism continues—and with plenty of adult support.
5. This film is hard to find, so I'm relying on Schaefer's summary.
6. *Risky Business* was directed by Paul Brickman, whose only other film to date is *Men Don't Leave* (1990). *Risky Business* grossed 63.5 million dollars upon its release.
7. Laura Mulvey has argued persuasively that the objectification of women is a defensive posture, one that betrays a priori anxiety about the integrity of manhood. She writes, "It is man's narcissistic fear of losing his own phallus . . . which causes shock at the sight of the female genitals and the subsequent fetishistic attempt to disguise or divert attention from them" (11). Perhaps that shock and fetishistic compensation, and not only prudery or anxiety over ratings, explain why teen sex films rarely show female genitals, as opposed to breasts? In this section, Mulvey is thinking about the fetishistic artwork of Allen Jones, but we could offer a similar catalogue of hyperphallic fetish objects in *Risky Business*, chief among them Daddy's Porsche. Putting aside debates about the use-value of the maternal phallus and other, perhaps retrograde, psychoanalytic terms, we might speculate further that the male gaze, though defensive in the sense of shoring up male identity against a presumed external threat, is constitutive of rather than reactive to "female sexuality," and particularly in film.
8. There is also the delightful film *Billy Elliot* (2000), directed by Stephen Daldry, in which a young boy from a coal-mining family in northern England falls in love with ballet. Daldry's film affirms the joy of dancing for boys, be they gay or straight.
9. *House Party* inspired two sequels, but no imitations, except perhaps *Friday* (1995), starring Ice Cube as a gentle teen-in-the-hood. Neither sequel is as charming and well made as the original film, in my view. *House Party 2* (1991) takes Kid to college, stressing the importance of higher education, and in *House Party 3* (1994) Kid finds both a wife and a career in music management. Films for and about African Americans increasingly target adults; there is no African American teen film to speak of, although new television networks (especially the WB) have tried to compensate. In mainstream cinema, an assimilationist model of multiculturalism prevails, not real respect for and pleasure in diversity.
10. Some such films were cautionary tales, showing the dire consequences of bad behavior; representative titles include *The Bottle and the Throttle*, *Highways of Agony*, *Last Date*, and *Narcotics: Pit of Despair*.
11. For a useful analysis of menstruation films in cultural context, see Martin, "Parody" and "Periods."
12. In a sense, *American Pie* updates *Weird Science* (1985), in which two teenage boys create a living sex doll by hooking up their computer to a Barbie doll.

13. Among the classmates who watch are the members of the pop group Blink 182, whose song "Mutt" is featured on the soundtrack.
14. When he came back into the room, Jim tried unsuccessfully to cover up the webcam; he's unaware that everyone's watching him.
15. One of the genre's most successful filmmakers, Russ Meyer, had close ties with *Playboy* and made such nudie classics as *The Immortal Mr. Teas* (1958), which, as Schaefer points out, wasn't so immoral even by 1950s standards (338).
16. There are a few significant exceptions, particularly *The Incredibly True Story of Two Girls in Love* (1995), *Boys Don't Cry* (1999), and *But I'm a Cheerleader!* (2000), all American-made.

Works Cited

American Pie. Dir. Paul Weitz. Perf. Jason Biggs, Eugene Levy, Shannon Elizabeth, Chris Kline, Alyson Hannigan. Universal Pictures, 1999.

Bernstein, Jonathan. *Pretty in Pink: The Golden Age of Teenage Movies*. New York: St. Martin's, 1997.

Dixon, Wheeler Winston. "'Fighting and Violence and Everything, That's Always Cool': Teen Films in the 1990s." *Film Genre 2000: New Critical Essays*. Ed. W. W. Dixon. Albany: State University of New York Press, 2000. 125–41.

Eberwein, Robert. *Sex Ed: Film, Video, and the Framework of Desire*. New Brunswick: Rutgers University Press, 1999.

Foucault, Michel. *The History of Sexuality, Volume I: An Introduction*. Trans. Robert Hurley. New York: Vintage, 1980.

House Party. Dir. Reginald Hudlin. Perf. Christopher Reid, Christopher Martin, Robin Harris, Martin Lawrence, Tisha Campell, A. J. Johnson. New Line Cinema, 1990.

Mabry, Rochelle. "My Generation? John Hughes's Teen Films as Adult Product." Unpublished essay, cited with permission.

Martin, Michelle. "Periods, Parody, and Polyphony: Fifty Years of Menstrual Education through Fiction and Film." *Children's Literature Association Quarterly* 22.1 (Spring 1997): 21–29.

——. "Postmodern Periods: Menstruation Media in the 1990s." *The Lion and the Unicorn* 23.3 (September 1999): 395–414.

Moran, Jeffrey P. *Teaching Sex: The Shaping of Adolescence in the Twentieth Century*. Cambridge: Harvard University Press, 2000.

Mulvey, Laura. *Visual and Other Pleasures*. Bloomington: Indiana University Press, 1989.

Risky Business. Dir. Paul Brickman. Perf. Tom Cruise, Rebecca De Mornay, Joe Pantoliano, Richard Masur, Bronson Pichot, Curtis Armstrong. Warner Brothers, 1983.

Schaefer, Eric. *Bold! Daring! Shocking! True! A History of Exploitation Films, 1919–1959*. Durham: Duke University Press, 1999.

Shy Guy. Educational collaborator, Alice Sowers, Ph.D. (Director, Family Life Institute, University of Oklahoma). Coronet Instructional Films, 1947.

Smith, Ken. *Mental Hygiene: Classroom Films, 1945–1970*. New York: Blast, 1999.

Chapter 6

Power and Repression/Repression and Power: Homosexuality in Subversive Picture Books and Conservative Youth Novels

Angela M. Salas

> A society can exist without children, but not for long. Children are a prerequisite for cultural reproduction over time, and childhood literally exists as the site of enculturation. Adults and social institutions are invested in ascribing meaning onto and into childhood in order to maintain social order, and the socialization of children negotiates not only behavior patterns, but also identity formation. . . . Childhood thus exists in complex mediation with the society around it.
>
> (Woodson 33)

Most sex education occurs outside the classroom. Clothing choices, family attitudes, playground negotiations, dinner conversation, and other aspects of contemporary culture all combine to teach children, even young ones, about socially acceptable gender roles and sexual relationships. The heterosexist milieu in which youngsters are raised is seldom considered didactic, except by outsiders to that milieu or by resistant insiders. Instead, much of the socialization children undergo seems obvious and neutral, particularly to the youngsters themselves. But books, particularly children's books, seem more obviously didactic than do clothing and popular culture, which is, in part, why books can provoke passionate responses, both of enthusiasm and of anger. At stake in discussions about the educative value of particular books are both the "behavior patterns" and the "identity formation" of which Stephani Etheridge Woodson writes in the passage that serves as the epigraph to my chapter. Opponents of "excessive" openness postulate that simply knowing about same-sex relationships, especially when these relationships are presented in a positive way, can make a child more likely to become homosexual.

For a student of literature, the very different quality of responses to recent American books addressing homosexuality is striking. Since their publication, Lesléa Newman's *Heather Has Two Mommies* (1989) and *Gloria Goes to Gay Pride* (1991), as well as Michael Willhoite's *Daddy's Roommate* (1990) and its sequel *Daddy's Wedding* (1996), have been among the more controversial books on school and public library shelves. These books have also been taken off those shelves, or moved to the adult section of public libraries. The 1992 inclusion of Newman's and Willhoite's works as optional texts in the Children of the Rainbow Curriculum in New York City caused a well-documented storm of criticism, which contributed, at least tangentially, to the dismissal of the school system's chancellor, Joseph Fernandez. In contrast, many young adult (YA) novels addressing homosexuality, such as Bette Greene's *The Drowning of Stephan Jones* (1991), Chris Crutcher's *Ironman* (1996), and Paula Boock's *Dare Truth or Promise* (1999), arouse very little attention, aside from laudatory comments about their authors' bravery and insight. Critics sometimes express the opinion that it is "about time" that gay, lesbian, and straight teens be able to read books dealing with homosexuality. Clearly, many people hold different ideas about the importance of picture books to children's behavior and identity formation than they have about YA novels. It seems there is such a thing as educating children too young; at least where homosexuality is concerned, early ignorance is adult bliss.

I have chosen to examine the picture books and novels named above for a number of reasons. First, all except *Heather* were published in the 1990s, and thus arise out of approximately the same cultural moment. Second, Willhoite's and Newman's books are as pioneering in the field of the picture book as was John Donovan's gay-themed YA novel *I'll Get There. It Better Be Worth the Trip.* in 1969. Also, between them, Newman and Willhoite employ both female and male protagonists, and depict both lesbian and gay households. As for selecting Greene's, Crutcher's, and Boock's books from among the many YA titles addressing homosexuality in the 1990s, they seem interesting for particular reasons. Crutcher's novel shows a young man discovering the demands of loyalty and friendship after learning that a respected teacher is gay. Bo, the narrator, is not gay, but in order to accept the nurturing and guidance of this mentor, he will need to face and purge his own prejudice. Boock's book, which emanates from New Zealand rather than from the United States, depicts a lesbian relationship and the conflicts brought to bear upon it. These two novels, published in the late 1990s, differ markedly from Greene's earlier novel, in which a young woman learns to stand against the crowd after the bashing death of a gay man. Taken together, these seven books reveal a variety of attitudes toward homosexuality, as well as a variety of didactic impulses.

Roberta Seelinger Trites's volume *Disturbing the Universe: Power and Repression in Adolescent Literature* (2000) suggests that the different quality of responses to children's and YA texts results from different perceptions of what such books are intended for, and of how they might influence their readers. Newman's and Willhoite's books posit families headed by same-sex couples as unremarkable, complicated only by the occasional bigotry of other people. In contrast, YA books paint more complex, and possibly more retrograde, pictures of life for homosexual people, or for those who know and care about people who are homosexual. Early in *Disturbing the Universe*, Trites posits,

> The chief characteristic that distinguishes adolescent literature from children's literature is the issue of how social power is deployed during the course of the narrative. In books that younger children read . . . much of the action focuses on one child who learns to feel more secure in the confines of her or his immediate environment, usually represented by family and home. Children's literature often affirms the child's sense of self and her or his personal power.
>
> But in the adolescent novel, protagonists must learn about the social forces that have made them what they are. They learn to negotiate the levels of power that exist in the myriad social institutions within which they must function, including family; school; the church; government; social constructions of sexuality, gender, race, class; and cultural mores surrounding death. (2–3)

The patterns that Trites observes in children's and youth literature are similar to the ones that Woodson identifies about childhood in general in the epigraph to this chapter. Further, they operate in the works I have singled out for this brief analysis. Newman's and Willhoite's picture books are conventional in that they are indeed concerned with "one child who learns to feel more secure in the confines of his or her immediate environment, usually represented by family and home." These books certainly also "[affirm] the child's sense of self and her or his personal power." But more radically, in conceiving, writing, and publishing these books, their creators critique the received wisdom that gay or lesbian life is tragic and fraught with danger, suggesting to child readers that strong families come in all forms, and that gay and lesbian people lead fulfilled, ordinary lives. As such, these books seem designed to serve as a remedy for certain forms of homophobia before children are particularly aware of stigmas against homosexuality.

One further way in which these books engage with power is this: the children of Willhoite's and Newman's books *are* young children (as are the books' intended readers), and are thus unlikely to have critical or complicated views of the adults they love. They have not yet hit the rebellious stage, when parents' attitudes and behaviors seem generally mortifying; instead, the children adore their gay and lesbian parents unreservedly. One might plausibly argue that within these books the "child's sense of self and her or his personal power" depends upon the child's security within the family, and upon children's unconditional acceptance of their parents and their lives. Books such as Newman's and Willhoite's might, in fact, have a salving effect upon those children who recognize that their same-sex parented families are "non-standard," and who are worried about what this means about their parents and about themselves. Such texts may be variously described as sex education, social engineering, and bibliotherapy.

In contrast, many YA novels show more complicated familial and power relations than do children's books. In fact, at least some of the parents in these novels are so unable to love and nurture their children that these youngsters must seek their parenting outside of the home. Bo, the protagonist of *Ironman*, renegotiates his relationship with the teacher who has been his most unfailing ally after he learns that the man is gay. The teacher, Lionel Serbouseck, represents a more satisfactory male role model than Bo's father, who seems to believe that humiliating and frustrating his son will turn the boy into a strong man. The owner of a sporting goods store, Bo's father goes so far as to sponsor Bo's chief competition, a relay team, in the local triathlon. Further, although Bo will complete all three legs of the triathlon against this trio, Lucas Brewster provides their cyclist with a five-thousand-dollar ultralight bicycle, hoping to ensure that, between fatigue and the limitations of his heavier bike, Bo will be defeated. Lionel Serbouseck, in contrast, treats Bo fairly and tries to help him channel his energies in productive ways. He brainstorms ways for Bo to better himself, and he always believes in the younger man's capacity to mature and grow. Still, the news that his teacher is gay rattles Bo, who decides to avoid him. With prodding from Mr. Nak, his anger-management teacher, who asserts, "I don't mind workin' with a man's anger, but I have a hard time workin' with a man who turns his back on his friends" (142), Bo finally apologizes to Serbouseck for his coldness and ingratitude. It is worth noting that Bo's is a decidedly adult decision, based upon loyalty and his code of conduct, rather than upon the abundant love that Heather holds for her two mommies.

Somewhat similarly, much sadder and slightly wiser after her boyfriend causes the death of the title character, Carla in Greene's novel rejects peer pressure and embraces her mother's progressive values. And in Boock's *Dare Truth or Promise*, one protagonist, Louie, learns to live with her mother's and her own homophobia, while the other, Willa, is soothed by the unswerving regard in which her mother holds her. In fact, the car crash that moves the novel to its crisis is occasioned by Louie's despair over her mother's appalled response to her lesbianism. Thus, as Trites notes, the YA novels' protagonists "learn to negotiate the levels of power that exist in myriad social institutions within which they must function"—and learn, as well, what happens when one does not negotiate successfully. Bo creates a nurturing web of relationships for himself to replace his unsatisfactory family; Carla experiences firsthand the disapprobation that small-minded culture and the courts may inflict upon those who are socially stigmatized; and Willa and Louie discover whom they can and cannot count on for support.

While novels of the 1990s that address homosexuality have lower death counts than do many of those that came before them,[1] they still often present homosexuality as a problem to be dealt with, and as an occasion for the personal growth of straight characters. Bo, Louie's mother, and Carla all learn from the mistakes they make in responding to another character's sexual orientation; thus one didactic impulse that the authors of these books reveal is their desire to teach readers to treat gay and lesbian people with respect. Still, while the more controversial picture books seem to resist heterosexist assumptions about gay life (as tragic or fraught with loneliness, victimization, and fear), some YA titles reinforce them, rendering them less threatening to the status quo than are the picture books.

The Purported Perils of Picture Books

Newman's picture books *Heather Has Two Mommies* and *Gloria Goes to Gay Pride*, like Willhoite's *Daddy's Roommate* and *Daddy's Wedding*, posit the unremarkable ordinariness of families headed by lesbian and gay parents. Heather's Mama Kate and Mama Jane give their daughter a loving home, complete with pets, walks in the park, picnics, and freshly baked cookies. When Heather goes to school and realizes that her family is distinctive, her teacher shows all the children the variety of families they come from. Some of Heather's classmates' families are mixed-race, while others are headed by a single parent or by an extended family network. In short, Heather's two mommies may not be "typical," but the very concept of "typical" has become outmoded and limiting for many families in this book. For her own part, Gloria sees at Gay Pride both the kinship of gay people and the high level of community support for families such as her own.

Willhoite's books, like Newman's, examine the ordinariness of Nick's family life. Nick's father, Daniel, has divorced Nick's mother (a common phenomenon) and started a home with Frank. The men love each other and Nick, and lead quiet, pleasant, ordinary lives. The overriding message of all three books is that love, and not the gender make-up of a family, is what is important. Newman's exuberant, incredibly didactic picture books have considerably less finesse than do Willhoite's, but both authors strive mightily to make the point to their young readers that families come in all forms, and that no specific configuration ought to be either valorized or condemned.

One productive way to frame this discussion of the differences in how picture books and youth novels are received by adults may be to see these differences as the result of clashing ideas about the pedagogical implications of the works. For as Peter Hollindale has noted, many discussions about literary quality in children's books, what he characterizes as discussions between "book people" and "child people," "reflect a much larger controversy about the chief purpose of education" (4). Despite Perry Nodelman's admonishment that obsession about pedagogy "denies children's literature its rightful place in the canon of literature worthy of serious analysis" (x), many people foreground the didactic possibilities of books for children, valuing their content and educative purposes far above their literary quality. Thus John Leo, in a 1992 editorial for *U.S. News and World Report*, asserts that *Heather Has Two Mommies* and *Daddy's Roommate* do not merely endorse pluralism about and tolerance of homosexuality, but instead teach acceptance of it (16). Newman's and Willhoite's gay-friendly books seem too patently ideological for Leo, who would prefer that children's books delight, rather than instruct, youngsters. It would appear that Leo would also prefer that children not be instructed about family structures and lifestyles of which he evidently disapproves.

In contrast to those who, like Leo, suggest that books can be *dulce* without being *utile*, or sweet but not instructive, critic John Stephens asserts that writing for children "belongs firmly within the domain of cultural practices which exist for the purpose of socializing their target audience" (8). For Stephens, children's literature exists to inculcate desirable values and attitudes in the child who is reading or being read to, and is therefore entirely didactic and ideological at all times. Readers, however, often do not notice the values and attitudes conveyed by books unless these values in some way conflict with their own suppositions. Stephens suggests that when one considers a book free of ideology it only means that one has "just read a book which precisely reflects those societal presuppositions which you yourself have learned to subscribe to, and which are therefore invisible" (50). Stephens's argument may help explain why people who either love or hate *Heather* or the *Daddy* books seem so baffled by the intensity of the response of people holding the opposite opinion. The books' message of acceptance seems either patently inappropriate or obviously correct.

If we accept Stephens's premise that all texts have at least some educative function, it seems reasonable to conclude that many of the adults who describe Newman's and Willhoite's work as parodies of political correctness find the books' politics incorrect. Such critics perceive that their views of homosexuality may be undermined by these picture books, and that young readers may thus be led to accept homosexuality, despite their parents' objections. If we limit ourselves to considering content-based rather than aesthetically oriented criticisms of these texts, we may hypothesize that the many virulent responses to Newman's and Willhoite's books further suggest fear that children who read these books may be "recruited" to a sexual lifestyle that is portrayed wholesomely in Newman's and Willhoite's work, but which the adult reviewers find troublesome. Thus, to their logic, the books under discussion challenge the family's authority to teach its children about sexual conduct and orientation.

To add insult to injury, these books do so precisely by invoking the power of the strong family. Child or adult, readers will note that the children in Willhoite's and Newman's books have parents in loving, happy, basically untroubled relationships. The children themselves are happy, thriving, and sure of their parents' devotion. Indeed, if Newman's books include potential strife, as seen (briefly) in *Gloria Goes to Gay Pride*, outsiders cause these problems; they do not arise within the family unit. In *Gloria*, Mama Rose, Mama Grace, and all their friends (both women and men) appear healthy, attractive, secure, and friendly. In contrast, the page introducing anti-gay protesters and their picket signs depicts one protester as a rumpled older man, looking weak and drained by the weight of his bigotry.

This health and happiness stands in stark contrast to portrayals of gays in such books as Newman's notably *un*controversial book *Too Far Away to Touch* (1995), illustrated by Catherine Stock. In *Too Far Away* *to Touch*, the protagonist Zoe's Uncle Leonard is a gay man dying of an AIDS-related illness. Uncle Leonard is illustrated as a gaunt, frail, tragic young man, whose appearance affirms popular stereotypes of how gay men look. Were Greene's doomed Stephan Jones to be illustrated, he might look very much like Uncle Leonard. Images of healthy lesbian parents with healthy gay friends and enough self-assurance to march in a gay-pride parade are more transgressive than are images of ill, fragile young men, whose sickness may serve as a warning against the "perils" of homosexual relations.[2]

Willhoite's books take a similar strategy to that of *Heather* and *Gloria*. In *Daddy's Roommate*, Nick's father Daniel has started a life with Frank. Nick's mother shows real affection for her ex-husband and encourages her son's close relationship with him. She reveals no concern that homosexuality is "catching," or that she ought to waste even a moment fretting about whether her son will grow up gay or straight. When Nick asks about his father's relationship with Frank, Mom tells him, "Being gay is just one more kind of love," thus reassuring him of her good feelings toward the relationship and modeling acceptance and generosity. Later, in *Daddy's Wedding*, Nick's mother happily attends Daniel's commitment ceremony and helps her son prepare for his role in that day's events. Indeed, the close and loving relationship between Nick and his father's partner, Frank, is portrayed in the most positive fashion. The two gay men are healthy, loving, and robust, unlike Uncle Leonard, or his counterpart in Mary Kate Jordan's *Losing Uncle Tim* (1989). Daniel and Frank have a future before them, and that future includes a loving, unstrained relationship with Nick.

In short, we see that in the four books by Newman and Willhoite upon which I am focusing, the gay and lesbian characters are not alien or exotic; they are secure and committed to their children and their partners. As such, these characters critique perceptions that homosexuality is anti-family, anti-child, and potentially tragic. Quite the opposite: the lesbian and gay parents in these books take their relationships and their child-rearing responsibilities seriously, and they handle them enthusiastically and well. The community support portrayed in *Gloria Goes to Gay Pride* (surely not all of the smiling, clapping folks who watch the parade go by are themselves homosexual?) also undercuts stereotypical notions about gays and lesbians living in tiny, isolated subcultures, shunned by straight and presumably narrow-minded neighbors.

What, then, might Newman's and Willhoite's books teach youngsters about homosexuality? For one, in all four books, love trumps sex. The fact that same-sex partners love each other is made much of;

however, there is little overt suggestion that these adults have sexual relations.[3] Newman's and Willhoite's books prudently take the sex out of homosexuality, replacing sexual relations with affection and devotion, what Kenneth Kidd has referred to as "a normalizing rhetoric of family values" (114), and thus suggesting that gays and lesbians are like heterosexuals in everything but their sexual orientation. In employing this "rhetoric of family values," Newman and Willhoite uphold recognizably mainstream virtues of fidelity and respect for one's partner, demystifying same-sex relationships and suggesting that they are no more *sexual* than are heterosexual ones. By formulating their narratives in terms of love and affection, Newman and Willhoite give their critics the responsibility of raising, or reading in, the issue of sexual activity.

Conversely, when Leo argues that these books (most particularly when associated with school curricula) move beyond tolerance and into advocacy of lesbian and gay parenthood, his true concern is evidently that positive portrayals of homosexuality will recruit children to homosexuality. Implicit in this argument is an assumption that books can seduce children in ways that "mainstream" culture cannot, and that children who might otherwise be "normal" can be made gay by such positive representations. One might note that children who grow up gay have not been made straight by the ubiquity of cultural assertions that heterosexuality is both more fulfilling and safer than homosexuality. Since we still do not know to what extent sexual preference is a cultural product and to what extent it is a genetically determined behavior, there is no calculating the degree to which sexually didactic texts such as Newman's and Willhoite's may wield real power.

Still, the illustrations in *Daddy's Roommate* and *Heather Has Two Mommies* might reinforce the frightening recruiting possibility for nervous readers. When children's-literature critic Elizabeth A. Ford discusses the difficulties that *Heather* has presented for some adult readers, she notes that the way illustrator Diane Souza has pictured Heather elicits discomfort. Mentioning that Heather is shirtless in one illustration and wearing "boy's" clothes in other pages, Ford asserts, "Heather provokes the fear that gay or lesbian parents will produce gay or lesbian children because her clothing, her features, her body signal *androgynous child*, not *boy* or *girl*" (129). Heather is happy with her two mommies, and is herself unfettered by sartorial markers of femininity; therefore, she is a sort of rogue child, a person whose sexuality is fluid rather than rigidly encoded. And since Heather as central child character can be presumed to engage the sympathy and admiration of at least some young readers, nervous adults may worry that she raises the possibility that a child may grow up to be healthy, secure, and sexually transgressive. For Willhoite's part, the little boy in the *Daddy* books is a mini-version of his gay father. In the illustrations, both father and son have curly hair, round cheeks, big eyes, and looks of cherubic innocence. By having the son resemble his father, Willhoite suggests that this child may be like his father in other ways as well. He may even grow up to be gay.

Thus the lesbian and gay parents in Newman's and Willhoite's books are pictured as sexually chaste, but the illustrations of the principal child characters may resonate with adult readers, who can see the children as nascent homosexuals. If attractive, happy children, raised by wholesome gays and lesbians, may themselves become homosexual, might the child reader who identifies with these children not grow to accept homosexuality, homosexuals, and the possibility that she is, herself, a lesbian? As Ford suggests, "it is the fear of what children may learn about their own sexual identities, not about the sexuality of adults around them, that makes these books controversial" (128).

In Contrast: Young Adult Novels

While certain picture books dealing with homosexuality distress a number of vocal readers with their simplicity and unalloyed good cheer, books written for a YA readership seem seldom to register on these same readers' radar. Some commentators have argued that such differences in reception occur because adults are uncomfortable thinking of young children as sexual beings, while movies, television, and advertising fetishize teen sexuality. James L. Kincaid and Anne Higonnet have both written at length about such constructions of what Higonnet describes as "an ideally innocent childhood" (8) and the lengths to which our culture goes to protect the sanctity of this ideal. Kincaid has offended many with his forceful suggestion that obsession with child molesters, kiddie porn, and sexually suggestive advertising is really a means for adults to meditate over the bodies and sexuality of youth, while convincing themselves that such interest is on behalf of children.[4]

It's certainly possible that part of the difference in reception between picture books and YA books has to do with this notion of child as sexually innocent and teenager as sexually experienced. If childhood is innocent in all matters sexual, books addressing sexuality, even the sexuality of adults, may seem to raise for discussion things that ought not be raised. An "ideally innocent childhood" may therefore be threatened by the premature introduction of sex and sexuality. Janice Irvine argues persuasively that socially conservative assaults upon the existence of sex education in schools are predicated upon the assumption that speaking of sex is equivalent to "doing it"—that "sex talk" is performative. Irvine suggests that "censorship efforts depend not only on a conviction that words can perform evil but that 'evil will be averted if only the words are not uttered'" (58). Sex cues, many quite explicit, exist in abundance at the checkout line of the grocery store; on billboards, television, and radio; and in video games. While these cues rarely meet with concerted objections from parents, any institutional (school or library) broaching of sexuality with young children often inspires efforts at censorship because it is viewed as both ideologically aggressive and sexually predatory.

In contrast, adolescents are widely assumed to be sexually active and in need of education about what it is they are, or soon will be, doing. Moreover, this education may well lack the affirmative qualities that critics such as Leo so object to in *Heather* and its fellows. Trites suggests that adolescent literature is "as often an ideological tool used to curb teenagers' libido as it is some sort of depiction of what adolescents' sexuality actually is. . . . Such novels tend to be heavy-handed in their moralism and demonstrate relatively clearly the effect of adult authors asserting authority over adolescent readers" (85). YA authors attempt to teach youth to control their sexuality, and when they take homosexuality as their subject, their works "share a tendency to address how teenagers are affected when they develop their sexuality [and are] oppressed because [they are homosexual]" (102). Homosexuality, then, becomes a problem to be handled, rather than "one more kind of love," as the mother in *Daddy's Roommate* explains to her son.

In addition, young adults usually select their own books. Even when a friend, relative, or librarian chooses the book for its consumer, the reader's experience with the text is presumably private, undertaken without intercession or the didactic intent assumed when such a book is given or read to a young child. Still, as Trites notes, books addressed to young adults have didactic purposes, much as children's books do; the difference is that the older reader has presumably chosen this didactic experience, while the child reader is seen as more passive, less capable of rejecting the lessons taught. Indeed, the preachments of the YA novel about homosexuality are overt; detecting them, and thus resisting them, requires no great sophistication on the reader's part. For example, Greene's novel *The Drowning of Stephan Jones* asks readers to confront the consequences of homophobia; Crutcher's *Ironman* makes clear that a gay teacher can easily be a mentor to his straight male students; Boock's *Dare Truth or Promise* argues for being true to one's sexual self.

The clear educational agendas at work teach YA readers about tolerance for sexual difference as well as about self-acceptance.

The didactic impulses of these books do not detract from their value to readers, particularly if we accept Stephens's premise that *all* literature is didactic at *all* times; they're no different in their desire to educate than are other books on other topics. But their particular brand of didacticism has been widely applauded, in contrast to the complaints about the "unsubtle" form of sex education that takes place in *Heather* (qtd. Ford 129). For example, Christine Jenkins notes, "Reading is one of the few ways for adolescents to gather information privately about [homosexuality]" (82), while her fellow bibliographers Martha Cornog and Timothy Perper argue that libraries need books by, about, and for lesbians and gays. Cornog and Perper assert, "Over and over again, gay and lesbian writers tell how, as young people, they went to the library looking for explanations of their feelings about gender and sex" (223), and indeed, scholar David Bergman, for one, has written eloquently about how his sense of what it means to be gay was formed by literary representations of homosexuality (21). Finally, like Jenkins, Cornog, and Perper, bibliographer Robert F. Williams has education in mind as he compiles lists of books about homosexuality that are suitable for YA readership. His hope is "to provide gay teenagers with factual information and the experiences of other gay teenagers . . . so they realize that they are not alone, and to provide readings to straight teenagers to help eliminate stereotypes about homosexuality and overcome prejudice" (13). This agenda seems very similar to that animating Newman's and Willhoite's books, but the didacticism of the picture books has aroused considerably more anxiety than that of the adolescent novels.

Still, as Lou Willet Staneck has suggested, "The dual but conflicting responsibility of socializing under the guise of entertainment has constrained some of the best adolescent authors" (21). It seems also to have established a particular set of conventions to which authors may feel obliged to adhere. Jan Goodman lists a number of problems with the YA novels examined for her 1983 article "Out of the Closet, But Paying the Price: Lesbian and Gay Characters in Children's Literature." A few of the "assumptions, stereotypes, and inaccuracies" Goodman notes (14–15) are as follows, with notations about their continuing relevance to the three YA novels from the 1990s under consideration here:

1) It is physically dangerous to be homosexual (*Stephan Jones*)
2) Homosexuals lead lonely lives (briefly seen in *Dare Truth or Promise*)

3) Homosexual adults can influence children to be homosexual (Bo's father holds to this assumption in *Ironman*)
4) Homosexuals are white and affluent (all the gay or lesbian characters in these novels are either wealthy or white, or are both)[5]

Taking Goodman's cue, one might suggest that another reason (in addition to their readers' presumed ages) why YA novels evade controversy is that they sometimes reinforce, rather than critique, negative and stereotypical ideas about the lives of gay and lesbian people. The authors of these works may be trapped by the discourses surrounding homosexuality. As Michel Foucault famously asserted, there's no one discourse of power, embattled by one opposing discourse; rather, "discourses are tactical elements or blocks operating in the field of force relations" (101–02). Bergman extends Foucault, arguing that "since homosexuals have fashioned their sense of themselves out of and in response to the heterosexual discourse about them, homosexuality—even as conceived by homosexuals—cannot be viewed outside of the constructs of heterosexuality" (26). These constructs (homosexuality as other than, or deviant from, heterosexuality) are often negative and limiting, as Goodman notes in her list of assumptions and stereotypes, and their use takes much of the bite out of YA authors' work, regardless of how progressive these authors intend to be.

In particular, the ways in which homosexual characters are portrayed in many YA novels can be problematic. Indeed, in many YA novels, homosexual characters are still—as Goodman noted two decades ago—depicted as neurotic and unhappy, victimized by others and let down by authority figures. Furthermore, they are often portrayed as emotionally and physically weak, living in depressed isolation from the rest of the world. Bergman has suggested that gay men's early conceptions of their own homosexuality are "far too often built upon homophobic foundations, the conception of homosexuality fostered by 'compulsory heterosexuality,' as Adrienne Rich has termed it" (5). Some books may, despite their good intentions, reflect rather than critique the homophobic foundations Bergman identifies.

For example, *The Drowning of Stephan Jones* is well meant, addressing the effects of homophobia and demonstrating to young readers that physically attractive, popular people can be vicious bigots. Yet Greene's novel also suggests to readers that Stephan's emotional sensitivity and physical delicacy make him a target for his teenaged harassers' attention, while his partner Frank Montgomery's more apparent masculinity frightens the boys away. In Stephan's first solo encounter with his future killers, the narrative uses such words as "begged," "pleaded," and "trembled" to describe his actions, noting that he hangs his head while speaking to the younger men, attempts to run away from them, screams loudly on the street, and later cries when he finds himself alone (65–69). Implied, but never stated, is the notion that Stephan actually attracts his tormentors by fitting the stereotype of effeminate gay man. Both the book's title and Greene's narrative structure announce Stephan Jones's death, and nothing in the narrative makes Stephan a sufficiently rounded character that readers will feel the human tragedy when he dies. As such, *The Drowning of Stephan Jones* is a melodrama addressed to straight readers about the perils of homophobia, rather than a novel about the possibilities of life as a homosexual person. Jones is a victim, sacrificed so that straight readers may learn a lesson about tolerance, rather than a character whose development might foster real growth and understanding in the minds of readers.

Furthermore, Greene's narrative reflects and reinforces many distressing stereotypes about homosexuality and homosexual people. At the novel's end, the only way Frank Montgomery can get anything resembling retribution for Jones's death is to imply, in the hearing of the press corps, that Andy Harris, the acquitted murderer, had been Jones's lover. Montgomery publicly brands Harris, the homophobic child of a homophobic parent, gay. He thus reinforces the very myths about homosexual promiscuity and kinky gay-on-gay murders that have fed Harris, his father, and almost everyone involved in the trial. Montgomery loses Stephan to bigotry; his own rage and despair make him tap into this bigotry to exact revenge for his loss. But Greene's novel does not criticize or question Montgomery's self-destructive choice; it simply renders it as the only thing Montgomery can do in a horrendous situation. This troubling revenge, as well as the various other permutations of homophobic violence in Greene's novel, cannot be said to offer any emotional sustenance to a young gay or lesbian reader, although they *might* move straight readers to be more decent to homosexual people.

In contrast to Greene's novel, Crutcher's *Ironman* is not really "about" homosexuality, but it handles the subject with more nuance than does *Stephan Jones*. Crutcher makes Lionel Serbouseck, the gay teacher, such an intriguing, well-drawn character that his interchanges with Bo seem very important, both to the narrative and as a measure of Bo's growth to adulthood. If Bo is to be the likable, responsible man he is trying to become, he needs to confront and divest himself of his stereotypes about homosexuality and about the nature of any relationship he can have with a gay man. Another positive male figure in his life, Mr. Nak, takes Bo to task for his assumption that he needs to be careful of Mr. Serbouseck now that he knows the man is gay:[6]

> "Well, let me give you a little schoolin'," he said. "Heterosexuals wreak a heap more havoc in this world than homosexuals do. Hell, you got nothin' to be careful *of.* Teachers ain't allowed to mess with kids under any circumstances. It's got nothin' to do with sexual preference, it's got to do with age an' position. Lionel Serbouseck walks with as much integrity as any man I know. . . . You *have* to know that, Brewster. If you've got a brain in your head, you have to know that. Hell, you bitch about your daddy—all set in his ways—an' here you are, doin' the very same thing you bitch about most." (143)

Bo's decision to appreciate and respect Mr. Serbouseck—and to humbly beg the man's pardon—will resonate with readers who remember Huck Finn's decision to *go* to hell if that's what's necessary to be true to his friend Jim. "He told me some important things about fathers and sons that night," muses Bo, "and I trashed them when he told me the truth. Mr. Nak is right: shame on me" (145). The difficulty Bo has in sorting out his thoughts is dispiriting, yet this part of the narrative suggests a certain amount of bravery on Bo's part. An aggressively heterosexual teenage athlete, Bo will reject jock culture and his father's homophobia to do justice to a man who has risked much to help him become a good adult. In so doing, Bo learns that the stereotypes that trapped his thinking are outdated, incorrect, and deeply limiting. Rejecting bigotry not only makes Bo a "better" person, it also frees him to make his own decisions about all sorts of things.

Ironman is clearly didactic, most particularly on the subjects of sexism, loyalty, and responsibility. But it seems important to recognize the aspects of this novel that may render Crutcher's critique of heterosexism and homophobia safer, or at least less readily noticed. First, there is *Ironman*'s membership in a particularly "manly" genre, the sports novel. Moreover, this is a sports novel with an crowded didactic agenda. Bo learns tolerance of sexual difference in a book in which he learns many other things, including pride in his girlfriend's ferocity and competence when facing physical violence. Initially embarrassed that she has defended him against a bully, Bo learns that the "rules" that girls are weak and homosexual men are loathsome are simply incorrect. Thus *Ironman* is a novel of development, rather than a "problem novel"; as such, the lessons of tolerance may sit more easily with YA readers wary of being lectured to. Also, since the relationship between Bo and Serbouseck, particularly as regards Bo's attitudes about homosexuality, takes up relatively little space in the novel, despite being of decisive importance, the book may well be overlooked by conservative critics of gay-positive literature. Because the triathlon for which Bo is training, his troubled relationship with his father, and his lessons in self-control make the book look like a conventionally didactic novel about a young man's growth to maturity, few critics could accuse Crutcher of writing a book with a "pro-gay agenda" regarding homosexuality.

Somewhat similarly, in the case of *Dare Truth or Promise*, New Zealand author Boock both challenges and participates in discourses surrounding homosexuality as a problem to be dealt with. The book is radical in the ways in which it posits Willa and Louie's relationship as sincere, intense, and based upon mutual attraction and respect. Just as the picture books discussed earlier depend, ironically, on translating into a gay context "family values" that would look conservative in another setting, Boock's novel describes its heroines' mutual attraction and even their sexual activity in ways that make the narrative resemble YA novels of heterosexual love. Take the pivotal moment of Louie and Willa's first date:

> When Willa turned and kissed her, Louie thought in her head, *This is my first kiss.* It wasn't, of course, she'd kissed a number of boys, and done more too, but she'd never, ever felt as if she were falling off of a cliff. She'd never before felt as if her body were being turned to water from the inside out, or as if they were both whirling through space into an airless black vortex. Louie felt all these things, and above all, a disbelief, a wild, terrifying disbelief that this should be happening—no, not that she was in love with a girl, for it seemed suddenly absolutely natural that she should be in love with this girl—but that, god only knew how, this girl should love her back! (58–59)

The point here would seem to be not that kisses shared by lesbians are objectively more powerful than those shared by heterosexual couples, but that the standard rhetoric of teen romance (the loss of self, the emphasis on astonishment, the deliberately breathless sentence structure) may be applied to same-sex relationships.

Still, one wonders whether Louie's later botched suicide attempt, her cry for help against the homophobia that even she feels about her relationship with Willa, isn't consonant with the notion, identified by Goodman as a convention of gay or lesbian YA fiction, that being homosexual has a tragic aspect. While the novel is certainly daring in some ways, its resolution hinges on all the characters coming to their senses after Louie's car accident, an accident in which Willa "proves" her love by saving Louie from choking to death. From that standpoint, Boock's work reflects but does not undermine dominant assumptions about the quality of life homosexual people enjoy. And yet, the novel ends with the two lovers alive and together, willingly looking toward their future as a couple, despite the knowledge that they will meet more obstacles as they move forward with their lives together. This ending, in which they are stronger and wiser, represents a step forward from earlier YA novels about lesbian relationships, which were replete with violence, suicide, rape, and loss.[7]

Greene's novel is a problem novel, and Crutcher's is one that addresses homophobia in the context of many other issues. Boock's work is arguably the most interesting of the three novels, since it allows readers to see Louie and Willa interact instead of keeping homosexuality "decently" offstage. The two move from attraction, infatuation, and first kisses to furtive sexual relations, separation, and the decision that they really are meant to be together, despite the problems they will face. If Newman and Willhoite provide child readers with information about the ordinariness of loving same-sex relationships, Boock suggests that the quality of Willa and Louie's love is also "ordinary," inasmuch as they love each other and have had to go through hard times to commit to being together.

Certainly, the treatment of religion here may cause some readers to conclude that Boock is simply shifting the circumstances under which intolerance is permissible rather than calling for an end to prejudice of all kinds. As in Greene's novel and in other staples of the genre such as Nancy Garden's *Annie on My Mind* (1982), the most overtly religious characters (Louie's mother and Keith, the brother of Willa's first lover, Cathy) are both the meanest and the most homophobic. The Catholic Church takes a beating in the narrative until Boock switches tactics by having Louie's father and her priest call upon Christ's lessons of love to urge acceptance of the relationship. Fundamentalism, here embodied by Keith, is also a convenient scapegoat. Keith's horror at lesbianism, particularly in Cathy and Willa, upon whom he has a crush, is shared by the rest of Cathy's family. Traumatized, Cathy attempts suicide—symbolically inside her church, a church that condemns and frightens her. Boock's use of religion as a trope for bigotry allows readers to adapt Goodman's list of the assumptions and stereotypes of gay YA fiction to a new purpose, concluding that religious people are prone to violence directed toward themselves or others, are too insecure to be emotionally supportive even to family members, and are liable to "recruit" the young—with tragic results.[8]

Goodman suggests that many YA "problem" novels addressing homosexuality represent no real threat to the homophobic ethos of

Anglophone culture at the turn of the millennium. While socially incendiary picture books at least appear to threaten the foundation of homophobic discourse, adolescent novels often reinforce negative, frightening stereotypes about homosexuality even while they affirm that homosexuality is, of course, perfectly acceptable. Perhaps YA novels will catch up with children's picture books in due course, providing readers with more examples of young people successfully negotiating what Trites has identified as "the levels of power that exist in the myriad social institutions within which they must function" (3). If they do, their authors may attract more unfavorable attention than they do at present. Even so, it is unlikely that these authors and their books will ever attract the sort of ire that Newman and Willhoite have, since many critics see adolescents as less in need of protection from the sex-education lessons of gay-positive literature than are young children.

In a society so torn over the issue of homosexuality, books addressing this issue are bound to be a locus of controversy. It is interesting that picture books are more controversial than are YA novels, but this is understandable, given that children's and YA fiction are considered to have different didactic and socializing functions. The cheerfulness and optimism of certain books for young children, particularly when given the imprimatur of inclusion in school curricula and on library shelves, unnerve adults who fear that children's sexual identities and behaviors will be formed by what they have read. YA novels, in contrast, are self-chosen, read privately, and consumed by youth who are popularly, if incorrectly, presumed to be sexually active and sexually formed. As such, our standards about such novels' responsibility as tools for enculturation are much more relaxed than are the standards that we apply to picture books where the same question is concerned. Further, since YA novels have often been "problem" novels, replete with messages about the troubles lesbian and gay people face in a hostile society, they don't often risk being perceived as attracting youth to homosexuality. More recent YA novels have been more complex and more progressive in the worlds they imagine for gay, lesbian, and straight characters, but typically they are nonetheless not so radical as are such works as *Heather Has Two Mommies* and *Daddy's Wedding*.

Picture books and YA novels both serve many didactic functions. Radical or retrograde, they can reflect their readers' lives or enrich those readers' sense of life's possibilities. What all the works mentioned in this brief study have in common is a desire to demystify the lives of homosexual people and to encourage tolerance of sexual difference. These desires may be the source of social controversy, but they are also the source of sexual education. It will be interesting to see how and why such narratives evolve during this decade.

Notes

1. Aidan Chambers's *Dance on My Grave* (1982) comes to mind, as do Sandra Scoppetone's *Happy Endings Are All Alike* (1978) and *Trying Hard to Hear You* (1974). Writing about gay stereotypes in YA works, Michael Cart remarks that "many of these books perpetuate . . . the idea that to be homosexual is to doomed, either to a premature death or to a life of despair at the darkest margins of society" (1810). Dogs are run over, and there are a great many fatal car crashes in these books, thematically connecting (homo)sex and death.
2. For an interesting analysis of the phenomenon of the sick, and therefore safe, gay male in children's literature, see McRuer.
3. My son, Ian, was seven years old when he came across *Daddy's Roommate*. The look of surprise and "cognitive dissonance" that sprang to his face at the illustration of Daniel and Frank lying chastely in their shared bed was quite funny. Clearly, at seven, Ian already had an idea that mothers and fathers share their bed, but that other people do not. He wasn't thinking that the illustration implied sexual relations, but was certainly responding to that picture in terms of the norms of his lived experience.
4. Kincaid's *Child-Loving: The Erotic Child and Victorian Culture* (1992) and *Erotic Innocence: The Culture of Child Molesting* (1998) both cover this ground. In discussing whether children can be seduced to homosexuality by literature or gay-positive imagery, Simon Watney asserts similarly, "The narrative of 'homosexual promotion' should be regarded as a powerful fantasy which permits some heterosexuals to legitimately dwell on the image of children's bodies as objects of (homo) sexual desire" (399).
5. All the characters in these books I am considering are either White or wealthy, or both. Two YA novels that buck this stereotype and are worth scholarly attention are *From the Notebooks of Melanin Sun*, by Jacqueline Woodson (1995), and *Push*, by Sapphire (1996).
6. As a short Japanese American Texas cowboy who has lost his family to alcohol, Mr. Nak, a heterosexual, seems consciously constructed to disturb other kinds of stereotypes and preconceptions that readers may be harboring.
7. M. E. Kerr's *Deliver Us from Evie* (1994) is another exception to the grim, violent earlier generation of gay-themed YA novels. Kerr's novel features a young lesbian couple who leave their Arkansas community to be together. Despite their families' objections, and public disapprobation, they travel to New York and Paris, and are happy together. There is no indication that the cutting of family and cultural ties makes them feel exiled; instead, they have chosen each other, and the choice makes them happier than are the people they leave behind. Evie's brother, Parr, has a girlfriend, and Kerr portrays their relationship as shallow and unsatisfactory, particularly in comparison to Evie and Patsy's relationship. Parr is obsessed with fitting in, although he wants desperately to leave this farming community; his punishment for

"outing" his sister seems to be that he will fit in, and that he will be unable to leave home and the family farm.

8. These characterizations of religious fundamentalism might be explained, in part, by Boock's own memories of the role that such fundamentalism played in a pivotal moment in New Zealanders' attempts to have homosexual activity decriminalized. As journalist Neil Miller notes, "In 1985 and 1986, the country was absorbed by a rancorous effort to repeal the sodomy laws. Parliamentary maneuverings and debate continued for sixteen months; fundamentalist ministers from the United States flew in to lobby against the repeal. In the end, law reform was narrowly approved, forty-nine to forty-four. In the process, a conservative society, often compared to post-war Britain's, openly confronted the subject of homosexuality for the first time" (296–97).

Works Cited

Bergman, David. *Gaiety Transfigured: Gay Self-Representation in American Literature*. Madison: University of Wisconsin Press, 1991.

Boock, Paula. *Dare Truth or Promise*. New York: Houghton Mifflin, 1999.

Cart, Michael. "Saying No to Stereotypes." *Booklist* (1–15 June 1999): 1810–11.

Cornog, Martha, and Timothy Perper. *For Sex Education, See Librarian: A Guide to Issues and Resources*. Westport: Greenwood, 1996.

Crutcher, Chris. *Ironman*. New York: Greenwillow, 1995.

Ford, Elizabeth A. "H/Z: Why Lesléa Newman Makes Heather into Zoe." *Children's Literature Association Quarterly* 23.3 (Fall 1998): 128–33.

Foucault, Michel. *The History of Sexuality, Volume I: An Introduction*. Trans. Robert Hurley. 1978. New York: Vintage, 1980.

Goodman, Jan. "Out of the Closet, But Paying the Price: Lesbian and Gay Characters in Children's Literature." *Interracial Books for Children Bulletin* 14.3 (1983): 13–15.

Greene, Bette. *The Drowning of Stephan Jones*. New York: Bantam, 1991.

Higonnet, Anne. *Pictures of Innocence: The History and Crisis of Ideal Childhood*. New York: Thames and Hudson, 1998.

Hollindale, Peter. *Signs of Childness in Children's Books*. Stroud, Gloucestershire: Thimble, 1997.

Irvine, Janice. "Doing It with Words: Discourse and the Sex Education Culture Wars." *Critical Inquiry* (Autumn 2000): 58–76.

Jenkins, Christine. "Heartthrobs and Heartbreaks: A Guide to Young Adult Books with Gay Themes." *Outlook* (Fall 1988): 82–92.

Jordan, Mary Kate. *Losing Uncle Tim*. Illus. Judith Friedman. Niles, IL: Whitman, 1989.

Kidd, Kenneth. "Introduction: Lesbian/Gay Literature for Children and Young Adults." *Children's Literature Association Quarterly* 23.3 (Fall 1998): 114–19.

Kincaid, James R. *Child-Loving: The Erotic Child and Victorian Culture*. New York: Routledge, 1992.

——. *Erotic Innocence: The Culture of Child Molesting*. Durham: Duke University Press, 1998.

Leo, John. "Heather Has a Message." *U.S. News and World Report* (17 August 1992): 16.

McRuer, Robert. "Reading and Writing 'Immunity': Children and the Anti-Body." *Children's Literature Association Quarterly* 23.3 (Fall 1998): 134–42.

Miller, Neil. *Out in the World: Gay and Lesbian Life from Buenos Aires to Bangkok*. New York: Random House, 1992.

Newman, Lesléa. *Heather Has Two Mommies*. Illus. Diana Souza. Northampton, MA: In Other Words, 1989.

——. *Gloria Goes to Gay Pride*. Illus. Russell Crocker. Boston: Alyson Wonderland, 1991.

——. *Too Far Away to Touch*. Illus. Catherine Stock. New York: Clarion, 1995.

Nodelman, Perry. *Words About Pictures: The Narrative Art of Children's Picture Books*. Athens: University of Georgia Press, 1988.

Staneck, Lou Willet. "Adults and Adolescents: Ambivalence and Ambiguity." *School Library Journal* 55 (1974): 21–25.

Stephens, John. *Language and Ideology in Children's Fiction*. New York: Longman, 1992.

Trites, Roberta Seelinger. *Disturbing the Universe: Power and Repression in Adolescent Literature*. Iowa City: University of Iowa Press, 2000.

Watney, Simon. *Policing Desire: Pornography, AIDS and the Media*. Minneapolis: University of Minnesota Press, 1996.

Willhoite, Michael. *Daddy's Roommate*. Boston: Alyson Wonderland, 1990.

——. *Daddy's Wedding*. Boston: Alyson, 1996.

Williams, Robert F. "Gay and Lesbian Teenagers: A Reading Ladder for Students, Media Specialists and Parents." *ALAN Review* (Spring 1993): 12–17.

Woodson, Stephani Etheridge. "Mapping the Cultural Geography of Childhood or, Performing Monstrous Children." *Journal of American Culture* 22.4 (Winter 1999): 31–43.

Chapter 7
"No One Will Ever Know Your Secret!" Commercial Puberty Pamphlets for Girls from the 1940s to the 1990s

Michelle H. Martin

In 1976, I and my female classmates sat in a dark portable classroom, waiting breathlessly, as our adored Miss Tate, who taught us physical education, turned out the lights and started the 35mm film of Disney and Kimberly-Clark's 1947 *The Story of Menstruation*. As far as I knew, only one girl in my fifth-grade class had already gotten her period, and she was willing to answer our incessant questions, but now *Miss Tate* had taken on the job of teaching us about breasts and periods . . . and she didn't even seem embarrassed. We were in awe.

After the film and "the talk," we all trailed back to our classroom, hiding our aquamarine *Very Personally Yours* pamphlets so that the boys wouldn't see what we had gotten in P.E. Some of the girls may have been brave enough to ask the boys what *they* had done in P.E. that day, but I wasn't. When I got home, I studied *Very Personally Yours* carefully and decided that I would use the miniature calendar in the back to mark my periods when they came. Interestingly enough, though, the calendar wasn't the part of the booklet that stayed with me for years; it was the cartoon images of the girl who, in the film, takes such a cold shower that ice cubes come clanking down on her from the showerhead; the girl who vacuums under a chair by picking the chair up by one leg and raising it over her head, then letting it crash to the floor as she dissolves into tears; and the girl with no feet who rides her bicycle, head kerchief waving backward in the wind—all cautions about what *not* to do during menstruation. Although I had had a few cold showers at Girl Scout camp when my tent was last in line for the bathroom, after seeing

The Story of Menstruation and reading *Very Personally Yours*, I made sure that I paid close attention to water temperature when I menstruated. I never figured out what my period would *do* if I took a shower cold enough for ice cubes to pummel me, but I did know that I never wanted to turn the color of that kid in the movie.

School nurses and teachers who teach puberty education usually receive pamphlets like *Very Personally Yours* along with free sanitary product samples for girls and free deodorant samples for both girls and boys when they request educational films from companies such as Tambrands (makers of Tampax) or Procter & Gamble (makers of Always).[1] In chapter 3 in the present anthology, Lesley Hall indicates that companies such as these also worked their way into school puberty-education programs in England in the 1950s, although it seems that puberty education in British schools was less widespread (and perhaps still is) than in American schools. Sharyn Pearce's timeline for boys' puberty and sex-education manuals also begins in the 1950s, suggesting that this was an important era for emerging school puberty-education programs in many Anglophone countries. While my own chapter will focus on the situation in the United States, it is worth noting that the link between puberty education and consumerism that I will explore here is not an exclusively American phenomenon.

To be sure, many school administrators dislike the blatant advertising in sponsored films—yet the prohibitive cost of non-sponsored films makes these free ones the only viable option for many schools. At one time, the commercial educational booklets that came with the films might have been a child's first exposure to puberty information. Even today, although (as Angela Salas articulates in chapter 6 of this volume) many children have already learned much from the media, books, their friends, and sometimes their parents, these booklets still play an important role in puberty education, both because most American schools use them and because the booklets often contain the first "official" information that children receive about puberty. Furthermore, these texts are the only part of the presentations that children take home with them to keep; the films stay at school, and children use and throw away the complimentary products.

While these pamphlets cost children nothing, they are far from free. Corporations spend thousands of dollars on the production, marketing, and distribution of these materials annually. Given the crucial function that they serve in the sale of menstrual products, these pamphlets more than pay for themselves through the service they render the corporation: encouraging young consumers to establish loyalty to a product that they will need for approximately the next four decades. But what of the personal cost to young consumers?

Girls at the receiving end of pamphlets such as Kimberly-Clark's *Very Personally Yours* and *Julie's Story* (1983) share a position with the young readers whom Jacqueline Rose describes in *The Case of Peter Pan: The Impossibility of Children's Fiction* (1984). Likening the relationship between the adult author and the child reader to "a soliciting, a chase, or even a seduction," Rose sees the unequal power relationship between writer and reader as somewhat insidious and disingenuous (2). If this is true of children's literature—a genre historically grounded in seeking to enhance the moral development of the child and to present ideas and values that will be "good for" the child—it is presumably even more true of a genre that exists *only* for the sale of corporate products. Like the printed material in kids' meals at McDonald's and Burger King or the games and toys that come in cereal boxes, these pamphlets attempt to make a positive and memorable connection with young people who influence adult spending. Unlike Happy Meal paraphernalia, however, whose sole purpose is entertainment, these pamphlets serve an important educational function—and one, I might add, for which many parents who would rather not discuss puberty or sex with their children feel grateful. But in abdicating this educational responsibility to corporations whose major concern is their profit margin, parents may not realize the extra baggage that their daughters are bringing home along with their pamphlets and free pads. I submit that the conflict between education and consumerism inherent in these pamphlets can result in ideological freight that advances misinformation about the relationship between pubescent development and reproduction, encourages shame about menstruation, and attempts to convince young women to see menstruation as just as much of a consumerist event as a physiological one. And although the marketing of the products in these pamphlets has changed radically since the 1940s, many of the potentially damaging messages about body image that characterize the earlier texts have arrived quite intact in the twenty-first century.

Throughout this thematic exploration of sponsored menstruation pamphlets, I have created section headings taken from quotations from Eve Ensler's *The Vagina Monologues* (2001) as well as from a variety of puberty pamphlets. Ensler interviewed over 200 women from a wide range of backgrounds about their relationships with their vaginas, which she then shaped into her widely praised off-Broadway production. By contrasting lines from Ensler's interviews of real women with lines from the menstruation pamphlets whose major function is marketing,

I would like to emphasize the connections between the lived experiences of real women and the fabricated experiences of puberty pamphlet human "mascots." These juxtapositions should also reveal that puberty pamphlets may serve as one source of the negative attitudes that many women—particularly American women—have about their bodies.

"She didn't even know she was pregnant . . ."[2]

I took Miss Tate's puberty education class in the 1970s, but the vast majority of girls who receive puberty education in American public schools still do so in classes separate from boys. The primary justification for this practice is that separating boys from girls alleviates some of the embarrassment inherent in these discussions and makes children more willing to talk about the physical changes they are undergoing. Whether or not single-gender classes actually do make the children more comfortable is debatable, but at least they make teaching easier for the adult because of another time-honored tradition of puberty education: divorcing puberty education from sex education. Given that the reason the human body goes through puberty is to enable sexual reproduction, this seems like an odd disconnect. But because parents and school boards often demand that puberty education exclude discussions of intercourse and sexual reproduction, educators must honor this bizarre demarcation and, just as crucially, must use educational materials that do likewise.

Parents have become even more insistent on this division between sex education and puberty education as the age for the onset of sexual development has steadily decreased over the past century and a half. If women began to menstruate at twenty-five, or even at sixteen—as they did in the mid-nineteenth century—when they are more likely to possess the emotional and financial stability to have children, their guardians and educators would be much less ambivalent than they currently are about menarche and about the information that girls receive about it in elementary and middle school. At one time, the medical profession defined "precocious puberty" as development prior to age eight in girls and age nine in boys. In 1997, however, a study of 17,000 girls showed that "7% of white girls and 27% of African-American girls begin developing breasts or pubic hair by age seven" (Gorman 84); moreover, "at age 8, 48% of blacks and 15% of whites" show evidence of pubescent development, even though the average age of menarche for American girls is still twelve (Painter 1A). In light of these facts, how do educators and parents negotiate giving a second-grade girl enough information to help her cope with changes that are occurring in her body years before they occur in the bodies of many of her classmates, while withholding from her knowledge about sex that she won't necessarily need—and may not be mature enough to understand—until several years later?[3]

The sanitary products industry's answer to this dilemma has been to put as much distance as possible between puberty education and sex education for as long as possible. Even though the Disney and Kimberly-Clark pamphlet *Very Personally Yours*, initially published in 1946, had undergone revisions in 1952 and 1959, its 1961 revision—the version that I received—still contained no explanation of conception. At the end of the discussion of ovulation in the 1961 booklet, the narrator says:

> If the egg becomes fertilized while passing through the tube (as it must be to start to become a baby), it will tuck itself into the prepared lining of the uterus. There, during the early part of its development, it will be fed by the extra nourishment that has been collecting in the uterine cells.
>
> But most eggs travel through the tubes without being fertilized, so they come to the end of their life cycle. Now that there's no further need for the enriched lining of the uterus, it breaks down—and in a few days it flows through the uterus opening into the vagina and is discarded. This flow has no more significance than many excess materials that the body discards. Items like fingernails, or eyelashes, or hair—that fall out or are snipped off. This material is eliminated about once a month. And because the Latin word for month is "mensis," this fluid is called "the menstrual flow" . . . And that's the entire story. (5–6)

Well . . . not exactly. But it was the entire story that the sponsors conveyed to young readers. And puzzled students who formulate questions such as, "But what makes the fertilization *happen*?" will likely get one version or another of, "Go home and ask your mother!"

Even pamphlets from the 1980s and 1990s dance delicately around this issue of fertilization. The 1983 Tambrands booklet *Accent on You* offers the following explanation of the process: "An organ called the uterus or womb prepares a lining where a tiny female egg cell can develop into a baby, if fertilized by a male sperm cell. If the egg is not fertilized, pregnancy does not take place and the lining of the uterus is shed" (5). As in the Disney pamphlet, the passive-voice or impersonal verb constructions in "if fertilized," "is not fertilized," and "does not take place" ensure that a male agent stays out of this reproductive picture. Similarly, the 1996 Procter & Gamble pamphlet *Always Changing* effectively circumvents any illuminating details about this process: "It is in the fallopian tube that an egg can be fertilized if sperm from a male are present. If the egg and a sperm unite in the fallopian tube, the fertilized egg will begin to develop into a baby" (7). Ironically named *A Time for Answers*, Tambrands's 1994 puberty booklet for girls, which details both male and female and sexual development, probably raises more questions than it answers because it manages to explain erection and ejaculation with no mention of intercourse or even masturbation:

> Hardening of the penis is called an erection. This occurs when blood rushes to the many tiny blood vessels in the penis, making it stiff so that it stands away from the body. An erection happens when a boy is sexually excited, but it can also happen at other times. Boys are often embarrassed when this happens, but they can't help it. The erection soon goes away.
>
> Sometimes when a boy or man has an erection, the muscles of the penis suddenly contract, pushing semen through the urethra and out of the body. This is known as **ejaculation** and happens first during puberty. If a boy ejaculates during his sleep, it's called a wet dream. (44–45, emphasis Tambrands's)

The deliberate exclusion of human agency in this passage could easily lead a boy to believe that an ejaculation could accost him at any moment of his waking or sleeping pubescent life; no mention of the involvement of the hands—his or anyone else's—or any other body parts implies that his penis might need assistance in producing an ejaculation.

The segregation of puberty education from sex education goes along with another telling omission in these publications: with few exceptions, puberty pamphlets for younger readers systematically omit the clitoris both from discussions and from anatomical illustrations, while those designed for high school audiences, who are presumably mature enough to learn about intercourse, include the clitoris. This may relate to the clitoris's function, which Hall discusses (chapter 3 of this volume) in the context of early twentieth-century British sex guides and which Ensler so deftly articulates:

> The clitoris is pure in purpose. It is the only organ in the body designed purely for pleasure. The clitoris is simply a bundle of nerves: 8,000 nerve fibers to be precise. That's a higher concentration of nerve fibers than is found anywhere else in the body, including the fingertips, lips, tongue, and it is twice . . . twice . . . twice the number in the penis. (51)

Of the eighteen puberty pamphlets in my collection, only six mention the clitoris: Kimberly-Clark's four 1981 Life Cycle Library pamphlets label and define it,[4] Kimberly-Clark's 1983 *Julie's Story* includes it in the text and glossary but not in the illustrations, and Procter & Gamble's 1994 *Always Changing* labels it in the illustration but excludes it from the glossary. Only one pamphlet—*Your New Self-Discovery*, the Life Cycle Library booklet for older teen readers—mentions the function of the clitoris, lists it in the glossary, and mentions its role in intercourse. Above the external illustration of the vagina, the text describes the parts of the vulva:

> The protective outer parts to the reproductive system are referred to as the vulva. The vulva is comprised of the *labia majora*—two soft folds of flesh on either side of the genital area, covered with hair in adult women. Enclosed within the *labia majora* are two smaller folds, the *labia minora*. These labia, or lips, join together at the upper ends to form folds which partly shelter the clitoris—a small, very sensitive organ which is easily stimulated to sexual arousal.

Later, in the explanation of intercourse, the text describes orgasm:

> At the climax (orgasm) of the sexual act, the penis ejaculates semen. No similar ejaculation takes place in a woman.[5] Although she may usually experience orgasm, she may not experience this every time she has intercourse. Orgasms of men and women differ and don't always take place at the same time. Varying in intensity and duration, the female orgasm produces a pleasurable sensation, especially in the clitoris. At other times, a woman experiences a pleasant feeling of relaxation. (22)

Because the clitoris seems to be the organ upon which discussions of intercourse and therefore pleasure as well as reproduction turn, the presence or absence of the clitoris in these pamphlets more often than not signifies the sponsor's commitment to talking or not talking about the relationship of pubescent development to sex. Yet many children *need* to know that various kinds of fluid exchanges can result in the contraction of sexually transmitted diseases and that post-menarcheal intercourse can result in pregnancy. Lacking this knowledge, is it any wonder that eleven-year-olds get pregnant, never quite realizing how it happened?[6]

Parents' and school boards' objections to sex education in American public schools help to dictate this corporate promulgation of opaqueness about how pregnancy actually happens. But even if conscientious parents fill in for their daughters the gaps that sponsored puberty pamphlets and films leave about this mysterious reproductive process, girls also absorb implicit messages about their bodies from this discourse about a product that serves a physiological purpose that most Americans still don't discuss in polite company. As with the disjunction between sex education and puberty education, the corporate marketing agenda can come at a personal cost to girls on the receiving end of these documents.

"Feel comfortable with who you are!"[—NOT!][7]

Scared people would smell it. Scared they'd say I smelled like fish.
(Ensler, *The Vagina Monologues*)

A consistent theme throughout the puberty pamphlets from Kimberly-Clark (makers of Kotex products), Personal Products (makers of Always brand sanitary products and, earlier, Modess), and Tambrands (makers of Tampax) is that although menstruation is "perfectly natural," women must pay special attention to their body odor, particularly while menstruating. In offering their products as the answer to unpleasant odors, "leakage," and other menstruation-related inconveniences, these companies convey negativity toward menstruating and suggest that all evidence of this monthly cycle should remain hidden.

One of the earliest pamphlets that I have found is a 1944 publication from the Zonite Products Corporation: *Newer Knowledge of Feminine Hygiene*. This is not a puberty pamphlet per se, and as far as I know, it did not accompany a sponsored film and targeted women more than girls. But I use it here for a number of reasons. First, it illustrates and accentuates a rhetorical tactic that still exists—even though more subtly—in many sponsored film pamphlets: it attempts to convince the reader that vaginas in general and menstrual fluid in particular create a specific and challenging hygiene problem whose resolution requires a product. Second, the pamphlet clearly advocates a generational approach to education, as do many of the early puberty pamphlets: mothers should tell their daughters about this revolutionary new solution to an age-old problem, a problem that the daughters may not yet know they have. The narrator advises, "To combat the source of infection mentioned, every girl and woman should know exactly how simple it is to cleanse the vaginal area externally with a germicide every day and it is wise to supplement external care with periodic douches of germicidal strength" (5). On the final page of the Zonite pamphlet appears a mother using the product on her injured daughter's arm.

Three pages earlier, the text includes the following exhortation to mothers:

> *Teach your daughter* about the need for this daily external care. Impress her at an early age with the part it plays as a precaution against infection and offensive odor. Be sure she also understands—as should all women—that during and immediately following menstruation, it is necessary to be especially meticulous about external cleansing, as at this time odor is particularly noticeable and doubly objectionable.[8] (13, emphasis Zonite's)

Finally, in addition to whatever else this pamphlet does, it, like puberty pamphlets, educates women and girls about their reproductive system. Although the clitoris is conspicuously lacking in the text and diagrams, the pamphlet discusses the womb, urethra, vagina, vulva, and anus, and provides a labeled illustration of all of these organs. For 1940s women who were unfamiliar with these anatomical terms, this Zonite pamphlet would have served the same purpose as puberty-education booklets do of educating readers while simultaneously submerging them in its own particular ideological message about the female body. In this case, the message is "You stink; let us help you."

In addressing married women about the problem of "feminine odor," the Zonite pamphlet offers an overt example of how the advertising agenda can vilify the female body even without intending to do so—a dynamic still at work in contemporary puberty pamphlets. *Newer Knowledge of Modern Feminine Hygiene* opens by suggesting that through a woman's ignorance of her own anatomy and hygiene, she can single-handedly destroy her own marriage:

> Two of the greatest threats to any woman's personal charm come from causes you may not even suspect. Most women don't. Yet one is a most serious deodorization problem so many women have. And it is one of the most important reasons for education about intimate feminine care. The danger to happiness lies in the very fact that women themselves are not aware of the problem.
>
> An even greater threat to happiness (because it can affect health itself) is the danger of vaginal infection present every day, in every woman's life. (*Newer Knowledge* 2–3)

And, of course, the pamphlet offers the solution to both odor and infection. It warns, "Daily baths are often not enough!" (9). Since—the pamphlet claims—frequent scrubbing can irritate delicate vaginal tissues and since soap might not destroy the source of the odor and works more slowly than this new odor-destroying technology, women need Zonite. The narrator recommends Zonite for external care as well as for douching, then praises it also as the best household first-aid antiseptic available.

Appealing to male medical expertise in the black-and-white illustrations—of a doctor in a white lab coat, in surgery, and pointing to a wall chart with a baton—the booklet explains the origin of Zonite, a product of the First World War. According to the pamphlet, Dr. Alexis Carrel, a French American surgeon, and H. K. Dakin, an English chemist, invented the Carrel-Dakin Solution during the war, and it revolutionized wartime health care because of its ability to "kill the most virulent bacteria without destroying body tissues" (14). Because of its effectiveness, the same technology used to treat war wounds and burns was improved, bottled, and made widely available to American households. "And the solution has been found to have other properties, peculiar only to the Zonite type of antiseptic, which make it particularly useful for vaginal irrigation" (14), since "the vulvar and outer vaginal areas can be a most potent source of offensive odor if neglected" (8).

In this pamphlet, the advertiser not only plays on women's fears about their feminine hygiene and its effect on the people around them, but also relies on a subtext that women's effluvia are unclean and therefore prone to disease and infection. The narrator points out that because of the close proximity of the anus, urethra, and vagina, few women can avoid the hygiene pitfalls that cause odor and various forms of vaginitis. "In this area are many glands which continually excrete substances"; hence, "Doctors whose patients are mostly women, have found a lack of cleanliness in this area among the majority of cases" (8). Although the narrator attributes this failure to girls being taught in childhood that touching their genitals is immodest, the message throughout the booklet remains strong and consistent: women are to blame for not keeping themselves clean and "dainty."

A 1926 advertisement for Sterizol, a douching contemporary of Zonite, claims that the product is just as effective for cleaning the bathroom floor as it is for douching ("Sterizol Douche Ad").[9] To contemporary readers, this dual marketing strategy may seem sinister. Gynecologists have found substantial evidence to suggest that douching damages vaginal health, killing helpful bacteria while enabling harmful bacteria to proliferate, and contributing to sexually transmitted diseases, infertility, tubal pregnancies, and possibly cervical cancer (Rome n.p.). Yet despite the harm that may result from using these stringent douching chemicals, women continued (and continue) to use them and encourage their daughters to do likewise because of the success of these advertising campaigns in convincing women that regular douching is the only way to cure offensive vaginal odors.[10]

Zonite's connection between odor and advertising makes sense for a product that is designed specifically to cleanse and rid the vagina of what was then presumed to be harmful and odor-causing. But this same connection also surfaces in the advertisements for products intended not for deodorizing but for the absorption of menstrual fluid. In the 1940 Cellucotton Company[11] puberty pamphlet *As One Girl to Another*, the text warns:

> You should take special care about your personal daintiness at this particular time. During these days your whole system is operating a bit differently. Your perspiring equipment is working overtime, and in unexpected places. So, to make sure you don't offend, put on a fresh pad every three or four hours. And when you change, sprinkle some Quest (the Kotex deodorant powder) on your napkin. If you do this, not even your worst enemy can make catty remarks about you! For Quest *positively* destroys body odors. (10, emphasis Cellucotton's)

If, then, the pad alone fails to keep the adolescent smelling fresh, surely the addition of the Kotex powder will.[12]

Notably, Kotex uses this same strategy—what Beverly Havens and Ingrid Swenson, whose 1988 study examines feminine hygiene advertisements targeted at teen magazine readers, call a "hygienic crisis" (qtd. Merskin 948)—to market another feminine hygiene product to teens. *As One Girl to Another*'s section "And What About Tampons?" asserts that authorities have declared tampons unsafe for teens unless they consult a doctor because tampons can break the hymen. This belief was once so widespread that tampons for teens were initially marketed for "unmarried girls," emphasizing that these tampons were small enough not to perforate the hymen of a virgin (Delaney et al. 113).[13] If, though, the doctor approves of tampon use, "then you'll want to try Fibs (for Fibs are made by the same company that makes Kotex—so you know you can have confidence in them)" (16). Well chosen, the name "Fibs" emphasizes secrecy, assuring a girl that she can safely lie that she is not menstruating because this tampon will never give her away. Many of the more recent pamphlets emphasize that because odor results from the exposure of menstrual blood to air, users of tampons have much less trouble with odor than do those who use pads. In addition to marketing these products by convincing women that they should consider vaginas and menstruation synonymous with foul odors, many of these companies did and still do insist that menstruation ought to be a secret event. In the end, this ideology has the power to convince young women that prideful or celebratory attitudes toward their growing bodies are somehow wrong or immodest and that through embracing the messages of secrecy and shame that bombard them, they can better conform to social expectations of attractiveness and desirability.

"Your narrow Kotex belt won't show under the tightest dress!"[14]

> *Thirteen years old. It was before Kotex. Had to watch your dress. I was black and poor. Blood on the back of my dress in church. Didn't show, but I was guilty.*
> (Ensler, *The Vagina Monologues*)

In contemporary American television advertising, we see luscious mascara rolling onto long lashes, organic shampoo being massaged into the scalps of loudly orgasmic women, and bath gels splashing onto naked torsos under warm showers. Despite the centrality of sex in the ads, when we watch these commercials we know what these products are for: enhancing lashes, cleaning hair, and washing the body. In fact, for the advertiser to convince us of our need for these products, we *must* know what they are. Otherwise, we can't desire them. Therefore, excepting perhaps some recent prescription drug commercials with long strings of side effects attached at the end, commercials make the product's purpose clear. In ads for sanitary products, however, viewers hear about confidence and free lifestyles and see video clips of a hand pouring blue liquid onto a pad or putting a tampon into blue liquid to show its absorbency potential. Why do these companies obfuscate the fact that all of their products absorb blood? And absorb vaginal blood? I believe that this question lies at the heart of the problematic nature of advertising products associated with a part and process of the female body still considered too taboo to discuss with the American public. This discomfort with the vagina, with what comes out of it and what goes into it, is just as much a part of puberty-education discourses—both historical and contemporary—as it is a part of TV commercials.

In *The Curse: A Cultural History of Menstruation* (1976), Janice Delaney, Mary Jane Lupton, and Emily Toth comment on the pamphlets that they, as children, took home from school puberty film presentations in plain brown wrappers:

> We who grew up in the fifties think that these booklets were as important in shaping our permanent consciousness about menstruation, as the

> attitudes of our mothers and our friends. The message from the manufacturers, of course, was aimed at getting us to use more and more napkins, to send for "training kits" so we'd be prepared, paraphernalia-wise at least, for the big day (becoming a woman in the United States involves education for consumerhood as much as the kiddie shows and their paper cereals do). But the message behind the ads was hard-sell shame: *Cover it up*. Use our product because "nothing will show"; "no one will know"; "your secret will be safe." (94)

Contemporary Nike ads have gotten so slick that Nike now assumes that its name no longer needs to appear on its shoes. "Just Do It!" and the Nike swoop say it all, and the Nike Corporation displays its swoop on its line of clothing and footwear so that its consumers become walking billboards for Nike. Ironically, however, the message that sanitary products companies send to its consumers is "Buy our products and hide them." How would Nike fare with such a marketing approach? Not so well. But because of the difference in the nature of the feet and the vagina—both of which, commercials tell us, smell bad at times—Kimberly-Clark and Tambrands successfully market their products on the "buy and hide" principle.

The 1940 Cellucotton Products Company booklet *As One Girl to Another* makes explicit the message of hard-sell shame about which Delaney, Lupton, and Toth write:

> And don't think anybody can tell you're menstruating simply by looking at you. That's just a lot of nonsense! Anybody is apt to look a bit wilted on a hot day, or get circles under their eyes if they haven't slept well. You can go anywhere—to class, to the office, to parties—without being self-conscious and without fear that your gear will show! Your narrow Kotex belt won't show under the tightest dress. And neither will your Kotex pad. For Kotex has flat, *pressed ends* that never make tell-tale outlines . . . never give your secret away. (9, emphasis Cellucotton's)

This message shows up in much later pamphlets. *You're a Young Lady Now*, a 1961 revision of the original 1946 pamphlet accompanying Disney and Kimberly-Clark's film *The Story of Menstruation*, says emphatically, "NOTHING SHOWS, so no one else knows you are wearing a belt and napkin" (5), and again: "Naturally you want a napkin that doesn't show, and Kotex is constructed with flat pressed, tapered ends to assure against revealing outlines" (16). Twenty-two years later, in an Always booklet called *Changing: A Booklet for Girls*, the girl narrator tries on a panty liner at her mother's advice, and then a thin maxi: "It felt sort of funny at first, but pretty soon I practically forgot it was there.

And when I looked in the mirror, nothing showed. Not even from the back" (10). Even *Always Changing*, a 1996 booklet and one of the most recent in my collection, includes among its list of facts about menstrual protection: "Pads really can't be seen under most clothing. But ultra thin pads may help you feel sure of this" (18).

This message of the necessity of secrecy also surfaces in entertainment media for teen girls such as magazine advertisements. In her 1999 study of sanitary protection ads in *Teen* and *Seventeen* magazines between 1987 and 1997, Debra Merskin found that 78 percent of the ads played on the fears of young readers—particularly on the fear of bleeding through clothing—while 38 percent explicitly mentioned secrecy. And a total of 94 percent of the ads "focused specifically on being 'found out'" in class (951). If most puberty pamphlets are only thinly disguised ads, it stands to reason that these educational documents convey the same ideological messages as do these visual ads. But that puberty pamphlets educate girls about their bodies makes the presence of these messages here even more disturbing than they are in TV commercials and magazine ads.

"So why are there so many different kinds?"[15]

> *I was terrified. My mother showed me the thick sanitary napkins. I was to bring the used ones to the can under the kitchen sink. . . .*
>
> *I was so afraid. I started putting the used pads in brown paper bags in the dark storage places under the roof. . . .*
>
> *One night, I came home late and snuck into bed without turning on any lights. My mother had found the used pads and put them between the sheets of my bed.*
>
> (Ensler, *The Vagina Monologues*)

That few pubescent American girls or women can even think about menstruation without also thinking of disposable menstrual products—of whatever brand name—signals that manufacturers have succeeded in convincing girls and women that the consumption of these products is synonymous with menstruating. Furthermore, over the past six decades, manufacturers have striven to convince young female consumers that the consumption of menstrual products equals adulthood. If menstruation requires the use of disposable menstrual products, and the consumption of these products signals the arrival of adulthood for young women, then the sanitary products industry gives girls eager to enter into adulthood an incentive to buy, use, fetishize, and fantasize about their products. As Joan Jacobs Brumberg comments, female coming-of-age thus becomes "a process to be worked out in the marketplace rather than at home" (41).

Further, the more recent pamphlets make community an integral part of this consumerism. Since most contemporary films position girls within their daily social context (or at least a credible social context for a middle-class teen or preteen), the pamphlets reflect this trend. No longer does a disembodied, unnamed omniscient narrator tell the story about a solitary girl getting her period; now friends, sisters, mothers, and even fathers take an active role both in the puberty discourse and in the education of pubescent teens. As a result, pamphlets often convey a narrative with characters rather than teaching through a strictly instructional format. Typically the narrative features two or three friends, one of whom has gotten her period, the others of whom have not. The girl who has experienced menarche often gives her friends the *experiential* perspective (what it *feels* like to menstruate), while one of the girls who hasn't begun to menstruate but who reads widely fills in many of the information gaps for the menstruating girl and the other friends. In some of these pamphlets, such as Kimberly-Clark's 1983 *Julie's Story*, if the friend is just as misinformed as the menstruating girl, Mom and Big Sister step in to help.

Kotex's 1992 pamphlet *Becoming Aware: Sarah's Story* makes use of this narrative structure, in which the protagonists operate within a community of girls and women who can help them better understand their changing bodies. Sarah, the narrator, not only teaches Roxy, her friend who frets over telling her single-parent father that she has gotten her first period, but also creates a safe female community for Roxy by telling her mother about Roxy's menarche and by sharing her mother's Kotex with her, thereby postponing the need for Roxy to go to the store and buy her own—an embarrassing prospect for her. Roxy gets her first period at school and goes to Sarah for help. When Sarah asks her why she didn't go to the school nurse instead, she responds, "Because I have you" (15). Notably, many of the earlier pamphlets denigrated the advice that girls get from their friends; more and more, these relationships are affirmed as helpful and informative—even if only partially so.

When the discussion of products comes up, Sarah and Roxy conduct a covert search to uncover Sarah's mother's hidden stash of menstrual products on the top shelf of the "off-limits" closet. They plan to make a "variety pack" like the ones that cereals come in that would allow them to compare all the different sizes and brands and decide for themselves what they like best (15). When Sarah's mother comes home and catches them in her closet, she does not chastise the girls but offers to buy pads for Roxy to give her some time to adjust to buying them herself (or to convince her father to do so). Roxy readily accepts. This offer and Roxy's acceptance of it is a telling one. Even if the use of these products signals a girl's entry into womanhood on some level, buying them herself is an altogether different matter because it would require her public acknowledgment that she menstruates. Many teens avoid public consumption of these products (i.e., taking them to the checkout counter to a potentially male cashier, paying for them, and most likely carrying them out of the store in a semi-translucent plastic bag). Hence getting pads and tampons from other girls or women, or having others buy them for her, enables Roxy to forestall what she sees as public humiliation. Sarah comments, "I guess Mom knew Roxy wasn't real anxious to go fill a grocery cart with pads and tampons and all that stuff. With her luck, some great looking guy would be working at the cash register" (18). Sarah here points out an important paradox: while we admire girls who are assertive and self-directed, when it comes to taking care of pubescent well-being and needs, we often promote reserved behavior that might seem more appropriate to the shy, retiring, "dainty" teen of the Victorian era than to the contemporary American adolescent.

Magazine and television ads for feminine protection and hygiene products are commonplace now—though perhaps fraught with cultural expectations resembling those to which Roxy and Sarah are subject—but most teenagers still avoid buying these products themselves. My teenaged niece spends part of every summer with me, and in the summer of 2001, she told me that she had taught herself to use tampons while staying at my house. The day before she went back to Virginia, she left me the following note after I had gone to bed: "Dear Aunty Shelly, could you please go to the drugstore and buy me two boxes of Kotex Security Super Tampons? Thank you. Domonique." Having just worked on this section of the essay, I chuckled. But instead of granting her request, I waited until she woke up the next morning, took her to the drugstore with me, and had her pick what she wanted. She carried some of the boxes to the checkout counter herself and handed the clerk coupons for them that I had clipped from the Sunday paper. I asked her whether her mother buys sanitary supplies for her at home and whether teenagers are still as mortified to buy them themselves as I was as a teen. Without hesitation, she answered "Yes" to both questions. Ironically, then, while menarche signals a life stage into which many girls like Domonique enter with eagerness, and while public consumption of feminine hygiene products would confirm this level of maturation, the attitudes of shame and secrecy toward menarche and menstruation so integral to American popular culture and advertising and even to the documents designed to educate girls about their bodies strongly dissuade them from doing so.

Conclusion

Just as Rose describes children's literature as "an impossibility," perhaps "sponsored educational puberty pamphlet" is likewise something of an oxymoron. Perhaps asking corporations that exist only to make money off of consumers in a free marketplace to educate us well about their products and our need for them is always already an impossibility. Writer, educator, and counselor Lucinda Herring suggests in her essay "Celebration of the Maiden: Reclaiming the Beauty and Power of a Young Girl's First Moontime" that the very nature of *disposable* menstrual products speaks volumes about how these corporations view menses: as disposable and devoid of value and meaning. Herring writes:

> Modern day products do their best to keep us from a connection with our blood. We have all the most convenient ways to insure our periods are not obvious, do not slow us down or get in the way of more important endeavors. It took me a long time in my life to actually know and embrace my blood. The use of tampons makes it easy to just flush the telltale signs away and get on with your day. At least with pads or sponges you are seeing the blood as it falls from your body. (N.p.)

And *seeing* blood is the one visual that all commercial menstrual advertising—be it in television, magazines, or sponsored educational pamphlets—avoids. If "No one will ever know your secret," it's also true that no one will ever celebrate it. Mainstream American culture has not (yet) awakened to the possibility that menarche deserves celebration.

At the root of commercially sponsored puberty pamphlets' educational strategy of segregating sex education from puberty education and all that accompanies this practice ultimately lies the denial of the sexual power that girls gain along with menarche. Of this, Herring notes,

> When a girl bleeds she steps into her sexuality in a new way. Her womb opens and the ability to create life changes her forever. Yet she is unripe in every other way, ill prepared for the power germinating within her. Embracing such complexity is challenging. Add what it means to become sexual in the nineties and the task before us as parents becomes daunting. (N.p.)

Given the complexities surrounding puberty education in the early twenty-first century, it seems clear to me that only when Americans are ready to acknowledge a girl's sexual power and potential as a positive aspect of her sexual development at menarche will the divisions between sex education and puberty education begin to erode, and only then can more positive messages about the developing female body emerge.

Notes

I dedicate this essay to Mrs. Zelder Pressley, my high school freshman English teacher, who started me on this path to thinking critically about menstruation when she pulled me aside in English class one morning to tell me I had blood on my skirt when none of my female classmates had bothered. Though it turned out that I had sat in a puddle of blood on the school bus resulting from a violent fight between two boys, that incident got me thinking about what happens when women's bleeding goes public. And a special thanks to my Girl Scout buddy, Robin Campbell, for giving me pamphlets she found among her mother's papers.

1. In 1997, Procter & Gamble bought Tambrands for 1.85 million dollars, further narrowing the available options for menstrual products.
2. *Growing Up and Liking It: The Heather Report* 3.
3. In 1984, I worked at a Girl Scout camp in central South Carolina, and the waterfront staff had to pull a second grade girl out of the lake during swim time because she was menstruating and didn't understand that she couldn't wear a sanitary pad with her bathing suit. And who would teach an eight-year-old how to use a tampon? Dilemmas such as these will become more common as the age of menarche continues to plummet. While some researchers feel that pesticides in food and hormones in milk are responsible, new studies show that the rise in childhood obesity is more likely the cause of this phenomenon because fat cells produce a protein, leptin, which helps to trigger the onset of puberty (Gorman 84).
4. These titles include *The Miracle of You*, for elementary age readers; *Tell It Like It Is*, which details tampon use; *Very Personally Yours*, for middle school readers; and *Your New Self-Discovery*, the most extensive of the set.
5. Researchers John Delbert Perry and Beverly Whipple have found that women with strong pelvic muscle contractions do have the ability to ejaculate during intercourse (22).
6. According to 2001 Children's Defense Fund statistics, every thirty-seven minutes a baby is born to a mother who did not finish high school (Children's Defense Fund). I suspect that the lack of understanding that many girls have of how menstruation relates to reproduction contributes to this staggering figure; if so, this educational gap is costly both for children as individuals and for our society as a whole.
7. *Growing Up and Liking It: The Heather Report* 3.

8. In *The Vagina Monologues*, one of the questions Ensler asked was "What does your vagina smell like?" The responses she received from adults were variously positive and negative (ranging from "God" to "yummy candy" to "wet garbage"), but when Ensler asked a six-year-old girl what *her* vagina smelled like, she answered, "Snow flakes" (93–95, 104). I mention this as juxtaposition to Zonite's insistence that prepubescent girls must be as vigilant about fighting vaginal odor as are women.
9. Does this connection between the vagina and the bathroom floor suggest that both of these areas need not only to be "cleaned up" but also to be policed in some sense? If solutions such as Zonite and Sterizol could effectively keep infection at bay "in the field" during wartime, then—its male creators might have reasoned—how much more potent it could be in keeping unruly domestic germs in check. And perhaps the vagina and the bathroom floor both serve as appropriate metaphors for domesticity in need of controlling.
10. Much to my surprise, interested consumers can still purchase Zonite on the Shopinprivate.com website, even though the Museum of Menstruation describes this product as a vaginal irritant and potentially harmful ("Zonite").
11. Not wanting to associate itself with menstrual products at that time, Kimberly-Clark formed the Cellucotton Products Company in the late 1930s to market Fibs, a tampon, and Kotex pads. Likewise, Johnson & Johnson, best known for its baby products, is the parent company of Tambrands, but who knew? (Houppert 18). For decades, not wanting to sully their "No More Tears" image, Johnson & Johnson has kept their marketing of baby products separate from their marketing of sanitary supplies—another deliberate denial of any connection between a sexually mature "down there" and its role as a birth canal for those "No More Tears" babies.
12. Recent studies show a correlation between increased occurrence of ovarian carcinoma and the use of talcum powder in the vaginal area, although studies remain inconclusive as to how the frequency and duration of exposure affect the risk of this type of cancer (Chang and Risch 2396).
13. The objection to tampons was widespread. Catholic priests, for instance, feared that women would find tampons erotic and "worried that girls would lose their virginity upon insertion. (Their other concern: all those women and girls using their fingers to go exploring 'down there.' Who knows what they might learn along the way?)" (Houppert 16).
14. *As One Girl to Another* 9.
15. *Becoming Aware: Sarah's Story* 18.

Works Cited

Accent on You. Palmer: Tambrands, 1983.

Always Changing: A Booklet for Girls. Cincinnati: Procter & Gamble, 1994.

Always Changing: Guidebook for Girls about Growing Up. Cincinnati: Procter & Gamble, 1996.

As One Girl to Another. N.p.: International Cellucotton Products Company, 1940.

Becoming Aware: Sarah's Story. Neenah, WI: Kimberly-Clark, 1992.

Brumberg, Joan Jacobs. *The Body Project*. New York: Random House, 1997.

Chang, Stella, and Harvey A. Risch. "Perineal Talc Exposure and Risk of Ovarian Carcinoma." *Cancer* 79.12 (1997): 2396–401.

Changing: A Booklet for Girls. Cincinnati: Procter & Gamble, 1983.

Children's Defense Fund Website. Accessed 16 March 2002, www.childrensdefense.org.

Delaney, Janice, Mary Jane Lupton, and Emily Toth. *The Curse: A Cultural History of Menstruation*. New York: Dutton, 1976.

Ensler, Eve. *The Vagina Monologues*. New York: Villard, 2001.

Gorman, Christine. "Growing Pains: What Happens When Puberty Comes Too Soon in Your Child—and What You Can Do about It." *Time* (21 August 2000): 84.

Growing Up and Liking It: The Heather Report. Milltown: Personal Products, 1992.

Havens, Beverly, and Ingrid Swenson. "A Content Analysis of Educational Media about Menstruation." *Adolescence* 24.96 (1989): 901–08.

Herring, Lucinda. "Celebration of the Maiden: Reclaiming the Beauty and Power of a Young Girl's First Moontime." San Diego: Chinaberry, 1995.

Houppert, Karen. *The Curse: Confronting the Last Unmentionable Taboo: Menstruation*. New York: Farrar, Straus and Giroux, 1999.

Julie's Story. Neenah, WI: Kimberly-Clark, 1983.

Life Cycle Library. *The Miracle of You*. Neenah, WI: Kimberly-Clark, 1981.

———. *Tell It Like It Is*. Neenah, WI: Kimberly-Clark, 1981.

———. *Very Personally Yours*. Neenah, WI: Kimberly-Clark, 1981.

———. *Your New Self-Discovery*. Neenah, WI: Kimberly-Clark, 1981.

Merskin, Debra. "Adolescence, Advertising, and the Ideology of Menstruation." *Sex Roles: A Journal of Research* 40.11 (1999): 941–61.

Museum of Menstruation Website. Accessed 19 July 2001, www.mum.org.

Newer Knowledge of Modern Feminine Hygiene. New York: Zonite Products, 1944.

Painter, Kim. "Puberty Signs Evident in 7- and 8-Year-Old Girls." *USA Today* (8 April 1997): 1A.

Perry, John Delbert, and Beverly Whipple. "Pelvic Muscle Strength of Female Ejaculators: Evidence in Support of a New Theory of Orgasm." *The Journal of Sex Research* 17 (1981): 22–39.

Rome, Luci Capo. "The Perils of Vaginal Douching." Accessed 19 July 2001, www.mum.org/vagdouche.htm.

Rose, Jacqueline. *The Case of Peter Pan: The Impossibility of Children's Fiction*. Philadelphia: University of Pennsylvania Press, 1984.

"Sterizol Douche Ad." Accessed 19 July 2001, www.mum.org/Sterizol.htm.

The Story of Menstruation. Videocassette. Neenah, WI: Kimberly-Clark and Walt Disney, 1947.

A Time for Answers. Palmer: Tambrands, 1994.

Very Personally Yours. Neenah, WI: Kimberly-Clark, 1961.

You're a Young Lady Now. Neenah, WI: Kimberly-Clark, 1961.

Index

GPSR Compliance
The European Union's (EU) General Product Safety Regulation (GPSR) is a set of rules that requires consumer products to be safe and our obligations to ensure this.

If you have any concerns about our products, you can contact us on

ProductSafety@springernature.com

In case Publisher is established outside the EU, the EU authorized representative is:

Springer Nature Customer Service Center GmbH
Europaplatz 3
69115 Heidelberg, Germany

www.ingramcontent.com/pod-product-compliance
Lightning Source LLC
LaVergne TN
LVHW022220200926
848481LV00006B/2039
* 9 7 8 1 3 4 9 5 2 7 5 2 6 *